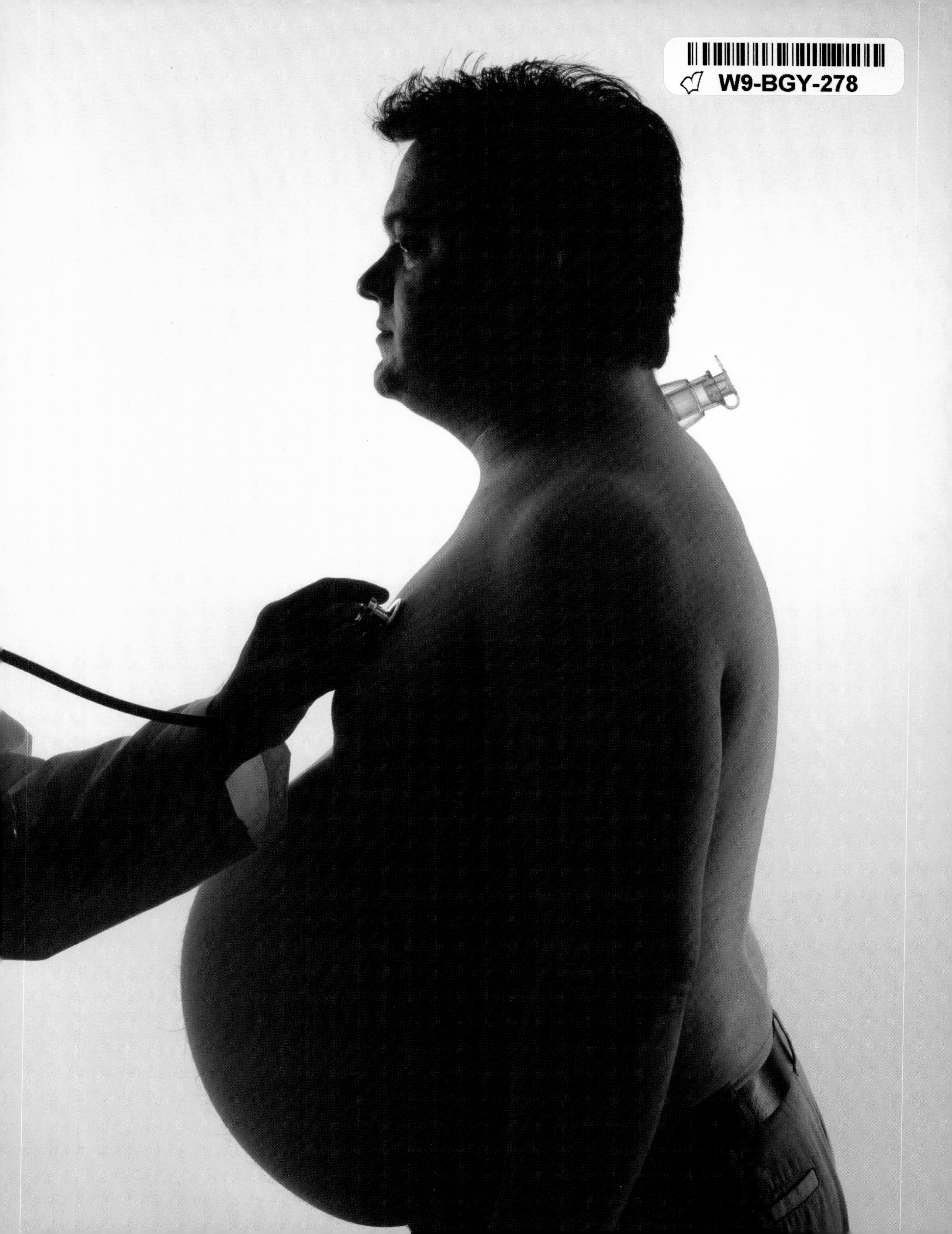

These critics, all academic researchers outside the medical community, do not dispute surveys that find the obese fraction of the population to have roughly doubled in the U.S. and many parts of Europe since 1980. And they acknowledge that obesity, especially in its extreme forms, does seem to be a factor in some illnesses and premature deaths.

They allege, however, that experts are blowing hot air when they warn that overweight and obesity are causing a massive, and worsening, health crisis. They scoff, for example, at the 2003 assertion by Julie L. Gerberding, director of the Centers for Disease Control and Prevention, that "if you looked at any epidemic—whether it's influenza or plague from the Middle Ages—they are not as serious as the epidemic of obesity in terms of the health impact on our country and our society." (An epidemic of influenza killed 40 million people worldwide between 1918 and 1919, including 675,000 in the U.S.)

What is really going on, asserts Oliver, a political scientist at the University of Chicago, is that "a relatively small group of scientists and doctors, many directly funded by the weight-loss industry, have created an arbitrary and unscientific definition of overweight and obesity. They have inflated claims and distorted statistics on the consequences of our growing weights, and they have largely ignored the complicated health realities associated with being fat."

One of those complicated realities, concurs Campos, a professor of law at the University of Colorado at Boulder, is the widely accepted evidence that genetic differences account for 50 to 80 percent of the variation in fatness within a population. Because no safe and widely practical methods have been shown to induce long-term loss of more than about 5 percent of body weight, Campos says, "health authorities are giving people advice—maintain a body mass index in the 'healthy weight' range—that is literally impossible for many of them to follow." Body mass index, or BMI, is a weight-to-height ratio [*see box on opposite page for the definition of weight categories*].

By exaggerating the risks of fat and the feasibility of weight loss, Campos and Oliver claim, the CDC, the U.S. Department of Health and Human Services and the World Health Organization inadvertently perpetuate stigma, encourage unbalanced diets and, perhaps, even exacerbate weight gain. "The most perverse irony is that we may be creating a disease simply by labeling it as such," Campos states.

## Overview/*A Crisis in Question*

- According to conventional wisdom, excess fat is an important cause of chronic disease, and the epidemic increase in obesity portends a coming health crisis.
- Four recent and forthcoming books by academic researchers argue that in fact the consequences of this trend for public health remain far less certain—and almost certainly less dire—than commonly suggested by obesity experts, government authorities and media reports.

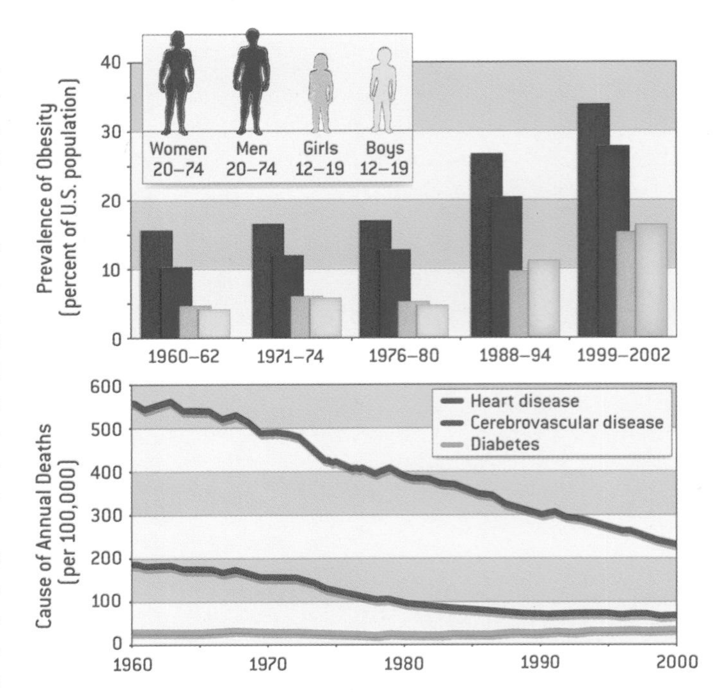


PREVALENCE OF OBESITY has roughly doubled in the U.S. since 1980 among adults and has tripled among children (*top*). Although deaths caused by diabetes have risen somewhat, predicted increases in mortality from heart disease and stroke have not materialized (*bottom*).

LUCY READING-IKKANDA; SOURCES: *HEALTH, UNITED STATES, 2004*, U.S. DEPARTMENT OF HEALTH AND HUMAN SERVICES; *STATISTICAL ABSTRACT OF THE UNITED STATES: 2004–2005*, U.S. CENSUS BUREAU

## A Body to Die For

ON FIRST HEARING, these dissenting arguments may sound like nonsense. "If you really look at the medical literature and think obesity isn't bad, I don't know what planet you are on," says James O. Hill, an obesity researcher at the University of Colorado Health Sciences Center. New dietary guidelines issued by the DHHS and the U.S. Department of Agriculture in January state confidently that "a high prevalence of overweight and obesity is of great public health concern because excess body fat leads to a higher risk for premature death, type 2 diabetes, hypertension, dyslipidemia [high cholesterol], cardiovascular disease, stroke, gall bladder disease, respiratory dysfunction, gout, osteoarthritis, and certain kinds of cancers." The clear implication is that any degree of overweight is dangerous and that a high BMI is not merely a marker of high risk but a cause.

"These supposed adverse health consequences of being 'overweight' are not only exaggerated but for the most part are simply fabricated," Campos alleges. Surprisingly, a careful look at recent epidemiological studies and clinical trials suggests that the critics, though perhaps overstating some of their accusations, may be onto something.

Oliver points to a new and unusually thorough analysis of three large, nationally representative surveys, for example, that found only a very slight—and statistically insignificant—increase in mortality among mildly obese people, as compared with those in the "healthy weight" category, after subtracting the effects of age, race, sex, smoking and alcohol consumption. The three surveys—medical measurements collected in the early 1970s, late 1970s and early 1990s, with subjects matched against death registries nine to 19 years later—indicate that it is much more likely that U.S. adults who fall in the overweight category have a *lower* risk of premature death than do those of so-called healthy weight. The overweight segment of the "epidemic of overweight and obesity" is more likely reducing death rates than boosting them. "The majority of Americans who weigh too much are in this category," Campos notes.

Counterintuitively, "underweight, even though it occurs in only a tiny fraction of the population, is actually associated with more excess deaths than class I obesity," says Katherine M. Flegal, a senior research scientist at the CDC. Flegal led the study, which appeared in the *Journal of the American Medical Association* on April 20 after undergoing four months of scrutiny by internal reviewers at the CDC and the National Cancer Institute and additional peer review by the journal.

These new results contradict two previous estimates that were the basis of the oft-repeated claim that obesity cuts short 300,000 or more lives a year in the U.S. There are good reasons to suspect, however, that both these earlier estimates were compromised by dubious assumptions, statistical errors and outdated measurements [*see box on page 7*].

When Flegal and her co-workers analyzed just the most recent survey, which measured heights and weights from 1988 to 1994 and deaths up to 2000, even severe obesity failed to show up as a statistically significant mortality risk. It seems probable, Flegal speculates, that in recent decades improvements in medical care have reduced the mortality level associated with obesity. That would square, she observes, with both the unbroken rise in life expectancies and the uninterrupted fall in death rates attributed to heart disease and stroke throughout the entire 25-year spike in obesity in the U.S.

But what about the warning by Olshansky and Allison that the toll from obesity is yet to be paid, in the form of two to five years of life lost? "These are just back-of-the-envelope, plausible scenarios," Allison hedges, when pressed. "We never meant for them to be portrayed as precise." Although most

## A Disease by Definition

**U.S. federal policy and WHO guidelines assign weight categories according to body mass index, or BMI, using the following formula and table:**

$$BMI = \frac{(\text{weight in kilograms})}{(\text{height in meters})^2}$$

| Below 18.5 | 18.5 to 24.9 | 25 to 29.9 | 30 to 34.9 | 35 to 39.9 | 40 or over |
|---|---|---|---|---|---|
| Underweight | Healthy weight | Overweight | Mild (class I) obesity | Moderate (class II) obesity | Severe (class III) obesity |

media reports jumped on the "two to five years" quote, very few mentioned that the paper offered no statistical analysis to back it up.

The life expectancy costs of obesity that Olshansky and his colleagues actually calculated were based on a handful of convenient, but false, presuppositions. First, they assumed that every obese American adult currently has a BMI of 30, or alternatively of 35—the upper and lower limits of the "mild obesity" range. They then compared that simplified picture of the U.S. with an imagined nation in which no adult has a BMI of more than 24—the upper limit of "healthy weight"—and in which underweight causes zero excess deaths.

To project death rates resulting from obesity, the study used risk data that are more than a decade old rather than the newer ratios Flegal included, which better reflect dramatically improved treatments for cardiovascular disease and diabetes. The authors further assumed not only that the old mortality risks have remained constant but also that future advances in medicine will have no effect whatsoever on the health risks of obesity.

If all these simplifications are reasonable, the March paper concluded, then the estimated hit to the average life expectancy of the U.S. population from its world-leading levels of obesity is four to nine months. ("Two to five years" was simply a gloomy guess of what could happen in "coming decades" if

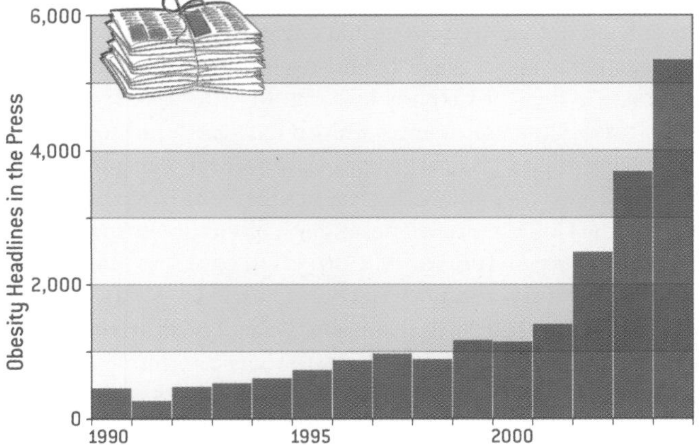

MEDIA STORIES on obesity are exploding in number, but Michael Gard and Jan Wright, authors of *The Obesity Epidemic: Science, Morality and Ideology*, charge the media with oversimplifying research results in ways that reinforce public prejudices and superstitions about body weight.

LUCY READING-IKKANDA; SOURCE: LEXIS-NEXIS, ADJUSTED FOR CHANGES IN DATABASE CONTENTS FOR 1990–2003

an increase in overweight children were to fuel additional spikes in adult obesity.) The study did not attempt to determine whether, given its many uncertainties, the number of months lost was reliably different from zero. Yet in multiple television and newspaper interviews about the study, coauthor David S. Ludwig evinced full confidence as he compared the effect of rising obesity rates to "a massive tsunami headed toward the United States."

**The overweight segment of the "epidemic of overweight and obesity" is more likely reducing death rates than boosting them.**

Critics decry episodes such as this one as egregious examples of a general bias in the obesity research community. Medical researchers tend to cast the expansion of waistlines as an impending disaster "because it inflates their stature and allows them to get more research grants. Government health agencies wield it as a rationale for their budget allocations," Oliver writes. (The National Institutes of Health increased its funding for obesity research by 10 percent in 2005, to $440 million.) "Weight-loss companies and surgeons employ it to get their services covered by insurance," he continues. "And the pharmaceutical industry uses it to justify new drugs."

"The war on fat," Campos concurs, "is really about making some of us rich." He points to the financial support that many influential obesity researchers receive from the drug and diet industries. Allison, a professor at the University of Alabama at Birmingham, discloses payments from 148 such companies, and Hill says he has consulted with some of them as well. (Federal policies prohibit Flegal and other CDC scientists from accepting nongovernmental wages.) None of the dissenting authors cites evidence of anything more than a potential conflict of interest, however.

## Is Fat Good for the Old?

"A lot of data suggest that the effect of obesity on mortality is less strong for old people than it is for young people," says Katherine M. Flegal of the CDC. "Some studies suggest that a high BMI is not a major risk factor among the elderly. Having a nutritional reserve seems to make people more resilient if they are hospitalized. So when you make estimates of deaths from obesity, it is very important which estimates you use for the oldest group. Obesity might be a tremendous risk factor in young people, but their death rates are very low."

RONNIE KAUFMAN Corbis

## Those Confounded Diseases

EVEN THE BEST mortality studies provide only a flawed and incomplete picture of the health consequences of the obesity epidemic, for three reasons. First, by counting all lives lost to obesity, the studies so far have ignored the fact that some diversity in human body size is normal and that every well-nourished population thus contains some obese people. The "epidemic" refers to a sudden increase in obesity, not its mere existence. A proper accounting of the epidemic's mortal cost would estimate only the number of lives cut short by whatever amount of obesity exceeds the norm.

Second, the analyses use body mass index as a convenient proxy for body fat. But BMI is not an especially reliable stand-in. And third, although everyone cares about mortality, it is not the only thing that we care about. Illness and quality of life matter a great deal, too.

All can agree that severe obesity greatly increases the risk of numerous diseases, but that form of obesity, in which BMI exceeds 40, affects only about one in 12 of the roughly 130 million American adults who set scales spinning above the "healthy" range. At issue is whether rising levels of overweight, or of mild to moderate obesity, are pulling up the national burden of heart disease, cancer and diabetes.

In the case of heart disease, the answer appears to be no—or at least not yet. U.S. health agencies do not collect annual figures on the incidence of cardiovascular disease, so researchers look instead for trends in mortality and risk factors, as measured in periodic surveys. Both have been falling.

Alongside Flegal's April paper in *JAMA* was another by Edward W. Gregg and his colleagues from the CDC that found that in the U.S. the prevalence of high blood pressure dropped by half between 1960 and 2000. High cholesterol followed the same trend—and both declined more steeply among the overweight and obese than among those of healthy weight. So although high blood pressure is still twice as common among the obese as it is among the lean, the paper notes that "obese persons now have better [cardiovascular disease] risk profiles than their leaner counterparts did 20 to 30 years ago."

The new findings reinforce those published in 2001 by a 10-year WHO study that examined 140,000 people in 38 cities on four continents. The investigators, led by Alun Evans of

## Mortal Mistakes

Media coverage of the obesity epidemic surged in 1999 following a report in the *Journal of the American Medical Association* by David B. Allison and others that laid about 300,000 annual deaths in the U.S. at the doorstep of obesity. The figure quickly acquired the status of fact in both the popular press and the scientific literature, despite extensive discussion in the paper of many uncertainties and potential biases in the approach that the authors used.

Like election polls, these estimates involve huge extrapolations from relatively small numbers of actual measurements. If the measurements—in this case of height, weight and death rates—are not accurate or are not representative of the population at large, then the estimate can be far off the mark. Allison drew statistics on the riskiness of high weights from six different studies. Three were based on self-reported heights and weights, which can make the overweight category look riskier than it really is (because heavy people tend to lie about their weight). Only one of the surveys was designed to reflect the actual composition of the U.S. population. But that survey, called NHANES I, was performed in the early 1970s, when heart disease was much more lethal than it is today. NHANES I also did not account as well for participants' smoking habits as later surveys did.

That matters because smoking has such a strong influence on mortality that any problem in subtracting its effects could distort the true mortal risks of obesity. Allison and his colleagues also used an incorrect formula to adjust for confounding variables, according to statisticians at the CDC and the National Cancer Institute.

Perhaps the most important limitation noted in the 1999 paper was its failure to allow the mortality risk associated with a high BMI to vary—in particular, to drop—as people get older [*see box on opposite page*].

Surprisingly, none of these problems was either mentioned or corrected in a March 2004 paper by CDC scientists, including the agency's director, that arrived at a higher estimate of 400,000 deaths using Allison's method, incorrect formula and all. Vocal criticism led to an internal investigation at the CDC; in January the authors published a "corrected" estimate of 365,000 obesity-related deaths a year, which they labeled as stemming from "poor diet and inactivity." The new figure corrected only data-entry mistakes, however.

Meanwhile another CDC scientist, Katherine M. Flegal, was preparing to publish a new and much improved estimate based entirely on nationally representative surveys that actually measured weights and heights. Flegal's analysis allows for risks that vary with age and claims to correct properly for confounding factors. But "the biggest reason that we get different results is that we used newer data," she asserts.

As illustrated in the chart below, the new analysis suggests that it is still far from certain whether there is any measurable mortality toll at all among overweight and obese Americans as a group. Even among the moderately and severely obese (those whose BMI exceeds 35), the plausible annual mortality found in the 1988–1994 survey ranges from 122,000 extra to 7,000 fewer deaths than one would expect based on the death rates of "healthy weight" people. —*W.W.G.*

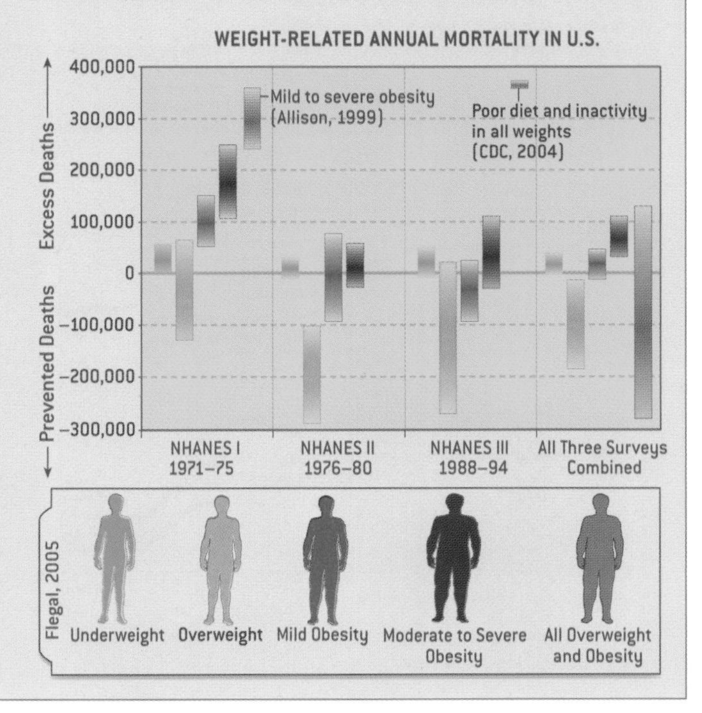


LUCY READING-IKKANDA; SOURCES: NATIONAL CENTER FOR HEALTH STATISTICS; JAMA, VOL. 282, PAGE 1530, OCTOBER 27, 1999, AND VOL. 291, PAGE 1238, MARCH 10, 2004

the Queen's University of Belfast, saw broad increases in BMI and equally broad declines in high blood pressure and high cholesterol. "These facts are hard to reconcile," they wrote.

It may be, Gregg suggests, that better diagnosis and treatment of high cholesterol and blood pressure have more than compensated for any increases from rising obesity. It could also be, he adds, that obese people are getting more exercise than they used to; regular physical activity is thought to be a powerful preventative against heart disease.

Oliver and Campos explore another possibility: that fatness is partially—or even merely—a visible marker of other factors that are more important but harder to perceive. Diet composition, physical fitness, stress levels, income, family history and the location of fat within the body are just a few of 100-odd "independent" risk factors for cardiovascular disease identified in the medical literature. The observational studies that link obesity to heart disease ignore nearly all of them and in doing so effectively assign their causal roles to obesity. "By the same criteria we are blaming obesity for heart disease," Oliver writes, "we could accuse smelly clothes, yellow teeth or bad breath for lung cancer instead of cigarettes."

As for cancer, a 2003 report on a 16-year study of 900,000 American adults found significantly increased death rates for several kinds of tumors among overweight or mildly obese

# Obesity and Illness

MILD AND MODERATE obesity seem in some studies to elevate risks of several serious diseases [*top chart*]. Yet the trends in these diseases [*middle* and *bottom charts*] reveal no simple connection between the epidemic rise in obesity and public health in the U.S.

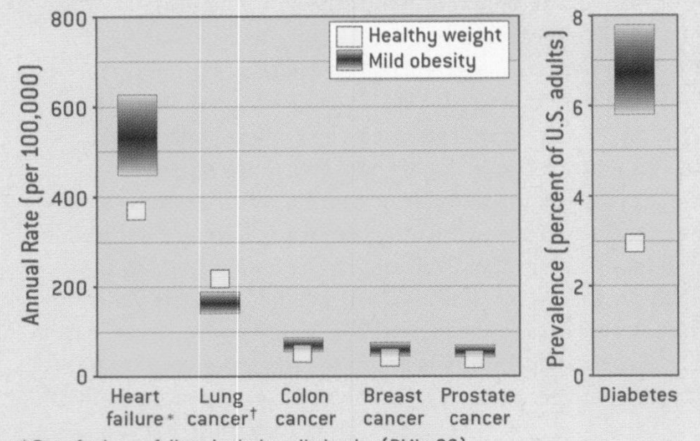


* Rate for heart failure includes all obesity (BMI >30)

† Rates for cancers represent mortality; lung and colon cancer in men only

DIABETES HAS RISEN along with obesity, but it did not spread significantly in the 1990s [*below, left*]. And major contributors to heart disease have fallen in recent decades [*below, right*].

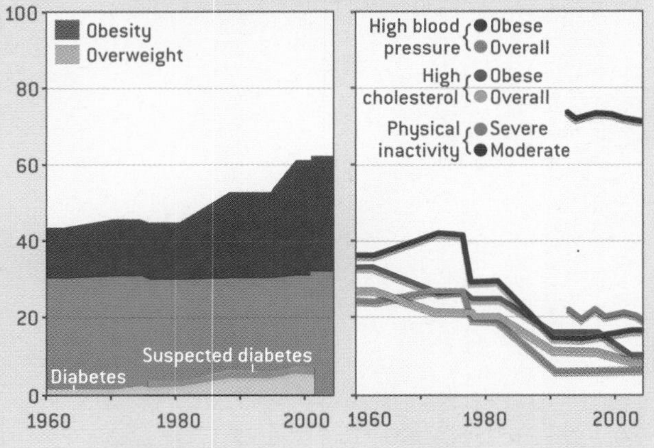


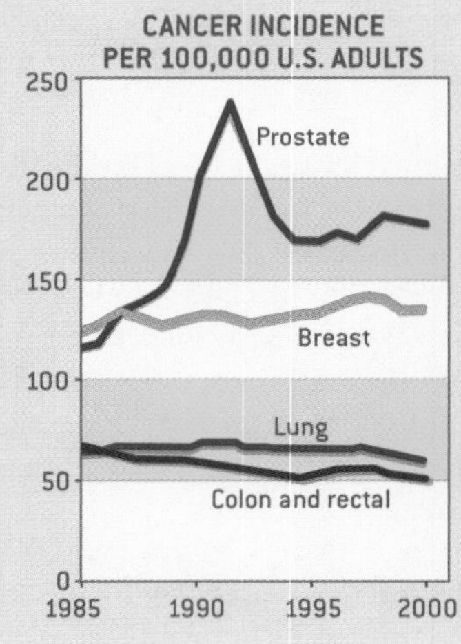


INCIDENCE OF CANCERS linked to obesity [*left*] also paints a complicated picture. New diagnoses of colon and of lung cancer have fallen slightly. (Fatness may actually protect against some lung cancer.) The upward trend in diagnosis of breast and prostate tumors may be a result mainly of increased screening for these diseases, as more sensitive and affordable tests catch tumors that previously escaped detection.

people. Most of these apparently obesity-related cancers are very rare, however, killing at most a few dozen people a year for every 100,000 study participants. Among women with a high BMI, both colon cancer and postmenopausal breast cancer risks were slightly elevated; for overweight and obese men, colon and prostate cancer presented the most common increased risks. For both women and men, though, being overweight or obese seemed to confer significant *protection* against lung cancer, which is by far the most commonly lethal malignancy. That relation held even after the effects of smoking were subtracted [*see box at left*].

## Obesity's Catch-22

IT IS THROUGH type 2 diabetes that obesity seems to pose the biggest threat to public health. Doctors have found biological connections between fat, insulin, and the high blood sugar levels that define the disease. The CDC estimates that 55 percent of adult diabetics are obese, significantly more than the 31 percent prevalence of obesity in the general population. And as obesity has become more common, so, too, has diabetes, suggesting that one may cause the other.

Yet the critics dispute claims that diabetes is soaring (even among children), that obesity is the cause, and that weight loss is the solution. A 2003 analysis by the CDC found that "the prevalence of diabetes, either diagnosed or undiagnosed, and of impaired fasting glucose did not appear to increase substantially during the 1990s," despite the sharp rise in obesity.

"Undiagnosed diabetes" refers to people who have a single positive test for high blood sugar in the CDC surveys. (Two or more positive results are required for a diagnosis of diabetes.) Gregg's paper in April reiterates the oft-repeated "fact" that for every five adults diagnosed with diabetes, there are three more diabetics who are undiagnosed. "Suspected diabetes" would be a better term, however, because the single test used by the CDC may be wildly unreliable.

In 2001 a French study of 5,400 men reported that 42 percent of the men who tested positive for diabetes using the CDC method turned out to be nondiabetic when checked by a "gold standard" test 30 months later. The false negative rate—true diabetics missed by the single blood test—was just 2 percent.

But consider the growing weights of children, Hill urges. "You're getting kids at 10 to 12 years of age developing type 2 diabetes. Two generations ago you never saw a kid with it."

Anecdotal evidence often misleads, Campos responds. He notes that when CDC researchers examined 2,867 adolescents in the NHANES survey of 1988 to 1994, they identified just four that had type 2 diabetes. A more focused study in 2003 looked at 710 "grossly obese" boys and girls ages six to 18 in Italy. These kids were the heaviest of the heavy, and more than half had a family history (and thus an inherited risk) of diabetes. Yet only one of the 710 had type 2 diabetes.

Nevertheless, as many as 4 percent of U.S. adults might have diabetes because of their obesity—if fat is in fact the most important cause of the disease. "But it may be that type 2 diabetes causes fatness," Campos argues. (Weight gain is a


LUCY READING-IKKANDA; SOURCES: *INTERNATIONAL JOURNAL OF OBESITY*, VOL. 26, PAGE 1050 (*heart failure*); *NEJM*, VOL. 348, PAGES 1625–1638 (*cancer mortality*); *JAMA*, VOL. 293, PAGE 1871 (*diabetes*); *JAMA*, VOL. 288, PAGE 1724, VOL. 291, PAGE 2847, AND *HEALTH, UNITED STATES, 2004* (*obesity and overweight*); *DIABETES CARE*, VOL. 27, PAGE 2809 (*diabetes trend*); *JAMA*, VOL. 293, PAGE 1871 (*blood pressure and cholesterol*); CDC BEHAVIORAL RISK FACTOR SURVEILLANCE SYSTEM (*physical activity*); NATIONAL CANCER INSTITUTE (*cancer incidence*)

DISTORTED VIEWS of medical research largely fuel the public's anxiety about the obesity epidemic, claims Paul Campos, author of *The Obesity Myth.* He castigates health authorities for a "constant barrage of scientifically baseless propaganda" about the risks of fat.

common side effect of many diabetes drugs.) "A third factor could cause both type 2 diabetes and fatness." Or it could be some complex combination of all these, he speculates.

Large, long-term experiments are the best way to test causality, because they can alter just one variable (such as weight) while holding constant other factors that could confound the results. Obesity researchers have conducted few of these so-called randomized, controlled trials. "We don't know what happens when you turn fat people into thin people," Campos says. "That is not some oversight; there is no known way to do it"—except surgeries that carry serious risks and side effects.

"About 75 percent of American adults are trying to lose or maintain weight at any given time," reports Ali H. Mokdad, chief of the CDC's behavioral surveillance branch. A report in February by Marketdata Enterprises estimated that in 2004, 71 million Americans were actively dieting and that the nation spent about $46 billion on weight-loss products and services.

Dieting has been rampant for many years, and bariatric surgeries have soared in number from 36,700 in 2000 to roughly 140,000 in 2004, according to Marketdata. Yet when Flegal and others examined the CDC's most recent follow-up survey in search of obese senior citizens who had dropped into a lower weight category, they found that just 6 percent of nonobese, older adults had been obese a decade earlier.

Campos argues that for many people, dieting is not merely ineffective but downright counterproductive. A large study of nurses by Harvard Medical School doctors reported last year that 39 percent of the women had dropped weight only to regain it; those women later grew to be 10 pounds heavier on average than women who did not lose weight.

Weight-loss advocates point to two trials that in 2001 showed a 58 percent reduction in the incidence of type 2 diabetes among people at high risk who ate better and exercised more. Participants lost little weight: an average of 2.7 kilograms after two years in one trial, 5.6 kilograms after three years in the other.

"People often say that these trials proved that weight loss prevents diabetes. They did no such thing," comments Steven N. Blair, an obesity researcher who heads the Cooper Institute in Dallas. Because the trials had no comparison group that simply ate a balanced diet and exercised without losing weight, they cannot rule out the possibility that the small drop in subjects' weights was simply a side effect. Indeed, one of the trial groups published a follow-up study in January that concluded that "at least 2.5 hours per week of walking for exercise during follow-up seemed to decrease the risk of diabetes by 63 to 69 percent, largely independent of dietary factors and BMI."

"H. L. Mencken once said that for every complex problem there is a simple solution—and it's wrong," Blair muses. "We have got to stop shouting from the rooftops that obesity is bad for you and that fat people are evil and weak-willed and that the world would be lovely if we all lost weight. We need to take a much more comprehensive view. But I don't see much evidence that that is happening." SA

*W. Wayt Gibbs is senior writer.*

## MORE TO EXPLORE


**Physical Activity in the Prevention of Type 2 Diabetes** in *Diabetes,* Vol. 54, pages 158–165; January 2005.

**Excess Deaths Associated with Overweight, Underweight and Obesity.** Katherine M. Flegal et al. in *Journal of the American Medical Association,* Vol. 293, pages 1861–1867; April 20, 2005.

**Secular Trends in Cardiovascular Disease Risk Factors according to Body Mass Index in US Adults.** Edward W. Gregg et al. in *Journal of the American Medical Association,* Vol. 293, pages 1868–1874; April 20, 2005.


MARC PoKEMPNER

# Obesity: An Overblown Epidemic?

*by W. Wayt Gibbs*

# IN REVIEW

ENDPOINTS

## TESTING YOUR COMPREHENSION

1) The prevalence of obesity has_____ in the U.S. since 1980.
   a) stayed relatively constant
   b) doubled
   c) quadrupled
   d) increased 10-fold

2) No safe and widely practical methods induce loss of more than about_____% of body weight.
   a) 5
   b) 10
   c) 15
   d) 25

3) Body mass index (BMI) is
   a) an imperfect indicator of body fat.
   b) a highly reliable indicator of body fat.
   c) a measure of a person's weight.
   d) the product of a person's weight and height.

4) Since 1960, the prevalence of heart disease has
   a) gone up with the increase in obesity.
   b) gone down with the decrease in obesity.
   c) gone up as the prevalence of obesity has gone down.
   d) gone down as the prevalence of obesity has gone up.

5) Compared to class I obesity, underweight is associated with
   a) excess deaths.
   b) the same number of deaths.
   c) fewer deaths.
   d) fewer deaths from heart disease but many more from diabetes.

6) A widely quoted figure in media reports with little scientific support is that obesity in the U.S. decreases average life expectancy
   a) one to two months.
   b) four to nine months.
   c) two to five years.
   d) eight to fifteen years.

7) The effect of obesity on mortality appears to
   a) increase with age.
   b) decrease with age.
   c) be independent of age.
   d) be highest for the very young and very old.

8) The disease with the clearest link to obesity is
   a) heart disease.
   b) stroke.
   c) high blood pressure.
   d) type II diabetes.

9) A recent study that directly examined people's heights and weights and corrected for age-related disease risks found that being obese
   a) carried the same risk as being underweight.
   b) substantially reduced mortality.
   c) substantially increased mortality.
   d) had no measurable effect on mortality.

10) The reason we don't know what happens when you turn fat people into thin people regarding their risk of early death and disease is that
   a) it's difficult to achieve this transformation.
   b) researchers who are largely funded by weight loss companies don't want to know.
   c) no one has wanted to make the effort to conduct such a study.
   d) bioethical review boards have denied approval of these studies.

# ENDPOINTS

## BIOLOGY IN SOCIETY

1) The idea that being overweight or obese is unhealthy has been branded into public consciousness. Recent studies discussed in this article call this accepted wisdom into question. Clearly, this issue is complicated and appears unresolved. How should government health agencies advise the public about weight and health? In your answer, you should consider the dual realities of uncertainty in the science surrounding the issue and how well the public deals with scientific uncertainty.

2) The article makes numerous references to obesity researchers receiving funds from weight loss companies. Is this necessarily a conflict of interest? Is this unethical? Should the practice be allowed to continue? If so, what mechanisms could be put into place to avoid the influence of corporate funding on the course and interpretation of obesity research?

3) If obesity is shown to have no significant adverse effects on health, how do you think this will influence the social stigma against excess weight? How large of an impact would this finding have on the weight loss industry? In general, how many of our strongest feelings are based on scientific truths?

## THINKING ABOUT SCIENCE

1) What is the body mass index (BMI) of a 5'10" man who weighs 225 pounds? What weight category is he in (see the table on page 5)? How much weight would this man have to lose to be at the upper end of the healthy weight range? What is your own BMI? Unit conversions for this calculation are 1" = 2.54 cm, 1 meter = 100 cm, and 1 pound = 0.454 kg.

2) One possibility cited in the article is that obesity is a marker, not a cause, of heart disease. If a correlation is established between obesity and heart disease, explain how it remains possible that obesity does not contribute to heart disease. What type of study could be done to learn if obesity does or does not directly contribute to heart disease?

3) Olshansky and Allison concluded that obesity in the U.S. currently shortens life expectancy by four to nine months. What assumptions did these investigators make in calculating the effect of obesity on life expectancy? Which of these assumptions, if any, are reasonable? What drives researchers to make such assumptions?

4) You are an investigator who asks how many early deaths in the U.S. are caused by the obesity epidemic. You know the percentage of the current population that is obese and the number of early deaths due to obesity. Is this information sufficient to provide an answer? If not, what other information do you need? How does the question targeted in your research differ from the question of how many early deaths in the U.S. are caused by obesity? Why is this subtle difference in phrasing the questions important?

## WRITING ABOUT SCIENCE

Your uncle is worried about the health effects of his weight and has asked you if it's worth the fight to try to lose weight. He's not clinically obese, but for him being "a bit thick around the waist" is something of an understatement. Write a letter to your uncle that lays out the current science on obesity's influence on health. Your letter should include information about the scientific controversy in this field and provide advice that will allow him to make an informed decision about whether he should work to lose weight. Since he has asked, offer him your opinion on what he should do.

**Testing Your Comprehension Answers:**
**1b, 2a, 3a, 4d, 5a, 6c, 7b, 8d, 9d, 10a.**

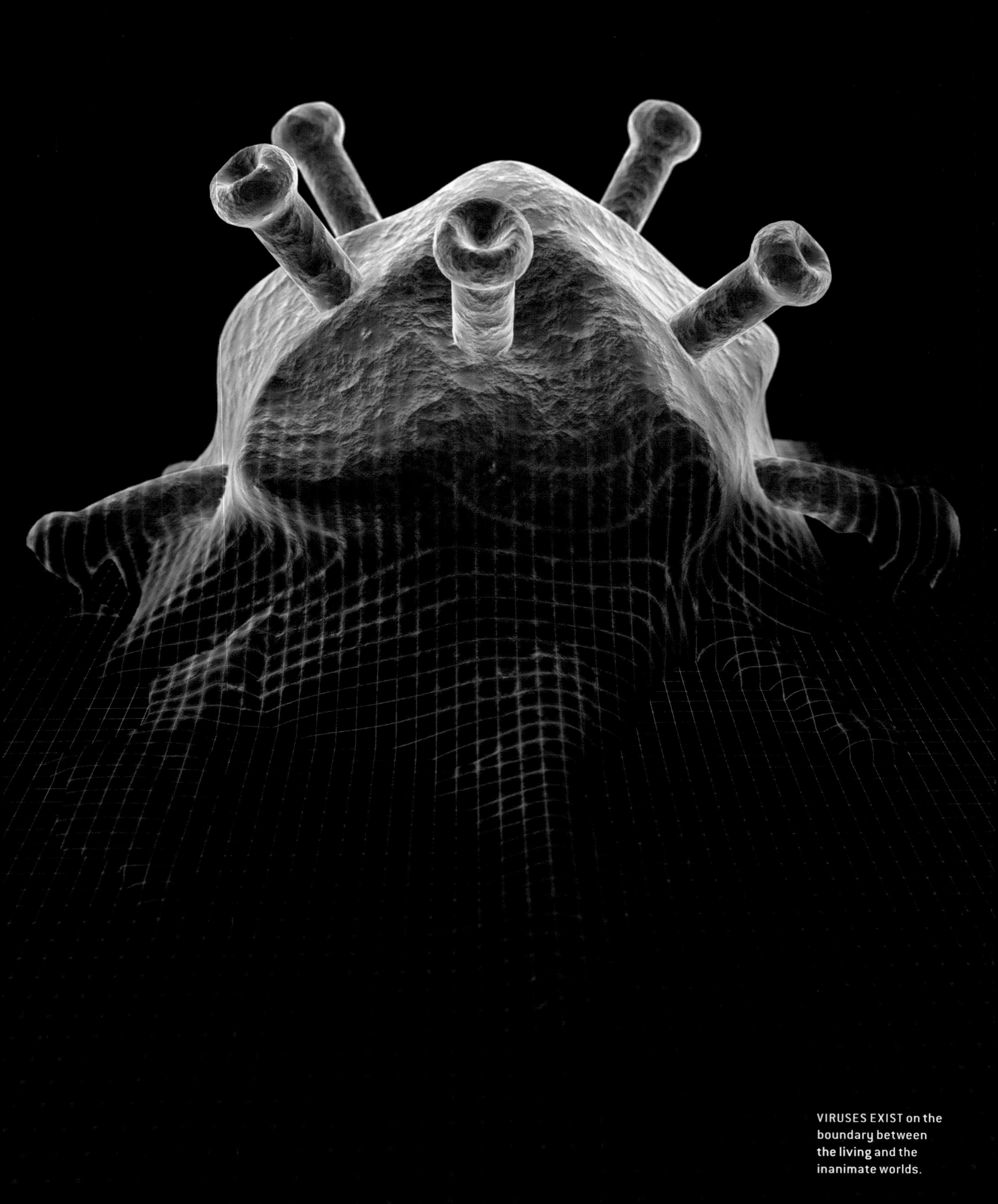

VIRUSES EXIST on the boundary between the living and the inanimate worlds.

BY LUIS P. VILLARREAL

# Are Viruses Alive?

**Although viruses challenge our concept of what "living" means, they are vital members of the web of life**

In an episode of the classic 1950s television comedy *The Honeymooners,* Brooklyn bus driver Ralph Kramden loudly explains to his wife, Alice, "You know that I know how easy you get the virus." Half a century ago even regular folks like the Kramdens had some knowledge of viruses—as microscopic bringers of disease. Yet it is almost certain that they did not know exactly what a virus was. They were, and are, not alone.

For about 100 years, the scientific community has repeatedly changed its collective mind over what viruses are. First seen as poisons, then as life-forms, then biological chemicals, viruses today are thought of as being in a gray area between living and nonliving: they cannot replicate on their own but can do so in truly living cells and can also affect the behavior of their hosts profoundly. The categorization of viruses as nonliving during much of the modern era of biological science has had an unintended consequence: it has led most researchers to ignore viruses in the study of evolution. Finally, however, scientists are beginning to appreciate viruses as fundamental players in the history of life.

## Coming to Terms

IT IS EASY TO SEE WHY VIRUSES have been difficult to pigeonhole. They seem to vary with each lens applied to examine them. The initial interest in viruses stemmed from their association with diseases—the word "virus" has its roots in the Latin term for "poison." In the late 19th century researchers realized that certain diseases, including rabies and foot-and-mouth, were caused by particles that seemed to behave like bacteria but were much smaller. Because they were clearly biological themselves and could be spread from one victim to another with obvious biological effects, viruses were then thought to be the simplest of all living, gene-bearing life-forms.

BRYAN CHRISTIE DESIGN

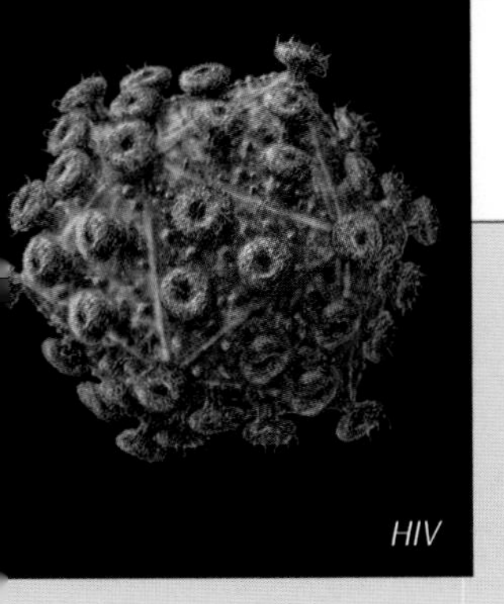
HIV

## What's in a Word?

"'Life' and 'living' are words that the scientist has borrowed from the plain man. The loan has worked satisfactorily until comparatively recently, for the scientist seldom cared and certainly never knew just what he meant by these words, nor for that matter did the plain man. Now, however, systems are being discovered and studied which are neither obviously living nor obviously dead, and it is necessary to define these words or else give up using them and coin others."

*—British virologist Norman Pirie, c. 1934*

"You think that life is nothing but not being stone dead."

*—George Bernard Shaw,* St. Joan, *1923*

Their demotion to inert chemicals came after 1935, when Wendell M. Stanley and his colleagues, at what is now the Rockefeller University in New York City, crystallized a virus—tobacco mosaic virus—for the first time. They saw that it consisted of a package of complex biochemicals. But it lacked essential systems necessary for metabolic functions, the biochemical activity of life. Stanley shared the 1946 Nobel Prize—in chemistry, not in physiology or medicine—for this work.

Further research by Stanley and others established that a virus consists of nucleic acids (DNA or RNA) enclosed in a protein coat that may also shelter viral proteins involved in infection. By that description, a virus seems more like a chemistry set than an organism. But when a virus enters a cell (called a host after infection), it is far from inactive. It sheds its coat, bares its genes and induces the cell's own replication machinery to reproduce the intruder's DNA or RNA and manufacture more viral protein based on the instructions in the viral nucleic acid. The newly created viral bits assemble and, voilà, more virus arises, which also may infect other cells.

## Overview/*A Little Bit of Life*

- Viruses are parasites that skirt the boundary between life and inert matter. They have the same kinds of protein and nucleic acid molecules that make up living cells but require the assistance of these cells to replicate and spread.
- For decades, researchers have argued over whether viruses are alive or not. This conflict has been a distraction from a more important issue: viruses are fundamentally important players in evolution.
- Huge numbers of viruses are constantly replicating and mutating. This process produces many new genes. An innovative gene, with a useful function, may on occasion be incorporated into the genome of a host cell and become a permanent part of that cell's genome.

These behaviors are what led many to think of viruses as existing at the border between chemistry and life. More poetically, virologists Marc H. V. van Regenmortel of the University of Strasbourg in France and Brian W. J. Mahy of the Centers for Disease Control and Prevention have recently said that with their dependence on host cells, viruses lead "a kind of borrowed life." Interestingly, even though biologists long favored the view that viruses were mere boxes of chemicals, they took advantage of viral activity in host cells to determine how nucleic acids code for proteins: indeed, modern molecular biology rests on a foundation of information gained through viruses.

Molecular biologists went on to crystallize most of the essential components of cells and are today accustomed to thinking about cellular constituents—for example, ribosomes, mitochondria, membranes, DNA and proteins—as either chemical machinery or the stuff that the machinery uses or produces. This exposure to multiple complex chemical structures that carry out the processes of life is probably a reason that most molecular biologists do not spend a lot of time puzzling over whether viruses are alive. For them, that exercise might seem equivalent to pondering whether those individual subcellular constituents are alive on their own. This myopic view allows them to see only how viruses co-opt cells or cause disease. The more sweeping question of viral contributions to the history of life on earth, which I will address shortly, remains for the most part unanswered and even unasked.

### To Be or Not to Be

THE SEEMINGLY SIMPLE QUESTION of whether or not viruses are alive, which my students often ask, has probably defied a simple answer all these years because it raises a fundamental issue: What exactly defines "life?" A precise scientific definition of life is an elusive thing, but most observers would agree that life includes certain qualities in addition to an ability to replicate. For example, a living entity is in a state bounded by birth and death. Living organisms also are thought to require a degree of biochemical autonomy, carrying on the metabolic activities that produce the molecules and energy needed to sustain the organism. This level of autonomy is essential to most definitions.

Viruses, however, parasitize essentially all biomolecular aspects of life. That is, they depend on the host cell for the raw materials and energy necessary for nucleic acid synthesis, protein synthesis, processing and transport, and all other biochemical activities that allow the virus to multiply and spread. One might then conclude that even though these processes come under viral direction, viruses are simply nonliving parasites of living metabolic systems. But a spectrum may exist between what is certainly alive and what is not.

A rock is not alive. A metabolically active sack, devoid of

RUSSELL KIGHTLEY *Science Photo Library*

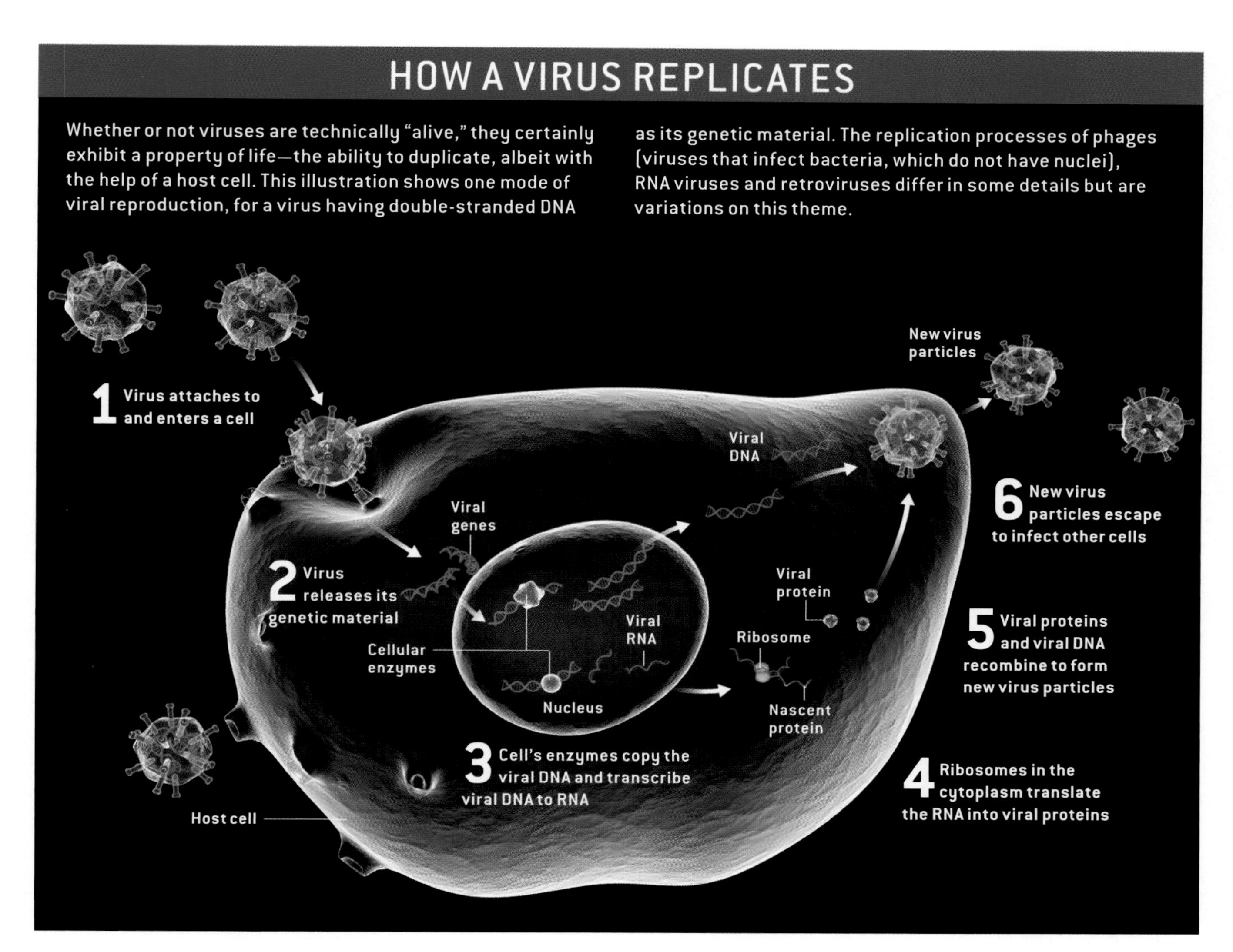


**HOW A VIRUS REPLICATES**

Whether or not viruses are technically "alive," they certainly exhibit a property of life—the ability to duplicate, albeit with the help of a host cell. This illustration shows one mode of viral reproduction, for a virus having double-stranded DNA as its genetic material. The replication processes of phages (viruses that infect bacteria, which do not have nuclei), RNA viruses and retroviruses differ in some details but are variations on this theme.

BRYAN CHRISTIE DESIGN

genetic material and the potential for propagation, is also not alive. A bacterium, though, is alive. Although it is a single cell, it can generate energy and the molecules needed to sustain itself, and it can reproduce. But what about a seed? A seed might not be considered alive. Yet it has a potential for life, and it may be destroyed. In this regard, viruses resemble seeds more than they do live cells. They have a certain potential, which can be snuffed out, but they do not attain the more autonomous state of life.

Another way to think about life is as an emergent property of a collection of certain nonliving things. Both life and consciousness are examples of emergent complex systems. They each require a critical level of complexity or interaction to achieve their respective states. A neuron by itself, or even in a network of nerves, is not conscious—whole brain complexity is needed. Yet even an intact human brain can be biologically alive but incapable of consciousness, or "brain-dead." Similarly, neither cellular nor viral individual genes or proteins are by themselves alive. The enucleated cell is akin to the state of being brain-dead, in that it lacks a full critical complexity. A virus, too, fails to reach a critical complexity. So life itself is an emergent, complex state, but it is made from the same fundamental, physical building blocks that constitute a virus. Approached from this perspective, viruses, though not fully alive, may be thought of as being more than inert matter: they verge on life.

In fact, in October 2004, French researchers announced findings that illustrate afresh just how close some viruses might come. Didier Raoult and his colleagues at the University of the Mediterranean in Marseille announced that they had sequenced the genome of the largest known virus, Mimivirus, which was discovered in 1992. The virus, about the same size as a small bacterium, infects amoebae.

**THE AUTHOR**

*LUIS P. VILLARREAL* is director of the Center for Virus Research at the University of California, Irvine. He was born in East Los Angeles. He received his doctorate in biology from the University of California, San Diego, and did postdoctoral research in virology at Stanford University with Nobel laureate Paul Berg. He is active in science education and has received a National Science Foundation Presidential Award for mentoring. In his current position, Villarreal has established programs for the rapid development of defenses against bioterrorism threats. He has two sons and enjoys motorcycles and Latin music.

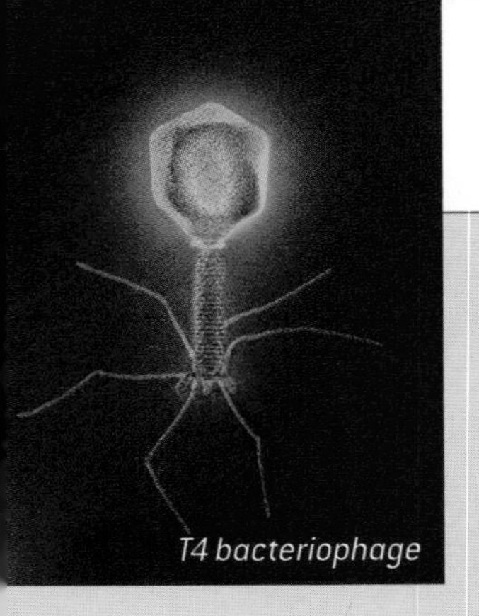
*T4 bacteriophage*

## Distracted by Cells

"Attention of biologists was distracted for nearly a century by arguments over whether viruses are organisms. The disagreement stems largely from the generalization put forth in the latter half of the nineteenth century that cells are the building blocks of all life. Viruses are simpler than cells, so, the logic goes, viruses cannot be living organisms. This viewpoint seems best dismissed as semantic dog wagging by the tails of dogma."

*—American evolutionary biologist Paul Ewald, 2000*

Sequence analysis of the virus revealed numerous genes previously thought to exist only in cellular organisms. Some of these genes are involved in making the proteins encoded by the viral DNA and may make it easier for Mimivirus to co-opt host cell replication systems. As the research team noted in its report in the journal *Science*, the enormous complexity of the Mimivirus's genetic complement "challenges the established frontier between viruses and parasitic cellular organisms."

### Impact on Evolution

DEBATES OVER WHETHER to label viruses as living lead naturally to another question: Is pondering the status of viruses as living or nonliving more than a philosophical exercise, the basis of a lively and heated rhetorical debate but with little real consequence? I think the issue *is* important, because how scientists regard this question influences their thinking about the mechanisms of evolution.

Viruses have their own, ancient evolutionary history, dating to the very origin of cellular life. For example, some viral-repair enzymes—which excise and resynthesize damaged DNA, mend oxygen radical damage, and so on [*see box below*]—are unique to certain viruses and have existed almost unchanged probably for billions of years.

Nevertheless, most evolutionary biologists hold that because viruses are not alive, they are unworthy of serious consideration when trying to understand evolution. They also look on viruses as coming from host genes that somehow escaped the host and acquired a protein coat. In this view, viruses are fugitive host genes that have degenerated into parasites. And with viruses thus dismissed from the web of life, important contributions they may have made to the origin of species and the maintenance of life may go unrecognized.

## Rising from the Dead—and Other Tricks

Because viruses occupy a netherworld between life and nonlife, they can pull off some remarkable feats. Consider, for instance, that although viruses ordinarily replicate only in living cells, they also have the capacity to multiply, or "grow," in dead cells and even to bring them back to life. Amazingly, some viruses can even spring back to their "borrowed life" after being destroyed.

A cell that has had its nuclear DNA destroyed is dead: the cell lacks the genetic instructions for making necessary proteins and for reproduction. But a virus may take advantage of the cellular machinery in the remaining cytoplasm to replicate. That is, it can induce the machinery to use the *virus*'s genes as a guide to assembling *viral* proteins and replicating the *viral* genome. This capacity of viruses to grow in a dead host is most apparent in their unicellular hosts, many of which live in the oceans. (Indeed, an almost unimaginable number of viruses exist on the earth. Current estimates hold that the oceans alone harbor some $10^{30}$ viral particles, either within cellular hosts or floating free.)

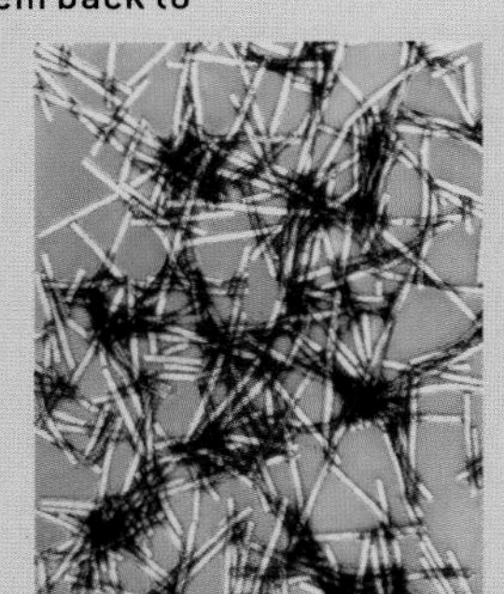
*Tobacco mosaic virus*

In the cases of bacteria, as well as photosynthetic cyanobacteria and algae, the hosts are often killed when ultraviolet (UV) radiation from the sun destroys their nuclear DNA. Some viruses include or encode enzymes that repair various host molecules, restoring the host to life. For instance, cyanobacteria contain an enzyme that functions as the photosynthetic center, but it can be destroyed by too much light. When this happens, the cell, unable to carry on photosynthesis and subsequent cellular metabolism, dies.

But viruses called cyanophages encode their own version of the bacterial photosynthesis enzyme—and the viral version is much more resistant to UV radiation. If these viruses infect a newly dead cell, the viral photosynthesis enzyme can take over for the host's lost one. Think of it as lifesaving gene therapy for a cell.

Enough UV light can also destroy cyanophages. In fact, UV inactivation is a common laboratory method used to destroy viruses. But such viruses can sometimes regain form and function. This resurrection comes about through a process known as multiplicity reactivation. If an individual cell harbors more then one disabled virus, the viral genome can literally reassemble from parts. (It is exactly such a reassembly capacity that allows us to create artificial recombinant viruses in the laboratory.) The various parts of the genome can also sometimes provide individual genes that act in concert (called complementation) to reestablish full function without necessarily re-forming a full or autonomous virus. Viruses are the only known biological entity with this kind of "phoenix phenotype"—the capacity to rise from their own ashes. *—L.P.V.*

(Indeed, only four of the 1,205 pages of the 2002 volume *The Encyclopedia of Evolution* are devoted to viruses.)

Of course, evolutionary biologists do not deny that viruses have had some role in evolution. But by viewing viruses as inanimate, these investigators place them in the same category of influences as, say, climate change. Such external influences select among individuals having varied, genetically controlled traits; those individuals most able to survive and thrive when faced with these challenges go on to reproduce most successfully and hence spread their genes to future generations.

But viruses directly exchange genetic information with living organisms—that is, within the web of life itself. A possible surprise to most physicians, and perhaps to most evolutionary biologists as well, is that most known viruses are persistent and innocuous, not pathogenic. They take up residence in cells, where they may remain dormant for long periods or take advantage of the cells' replication apparatus to reproduce at a slow and steady rate. These viruses have developed many clever ways to avoid detection by the host immune system—essentially every step in the immune process can be altered or controlled by various genes found in one virus or another.

Furthermore, a virus genome (the entire complement of DNA or RNA) can permanently colonize its host, adding viral genes to host lineages and ultimately becoming a critical part of the host species' genome. Viruses therefore surely have effects that are faster and more direct than those of external forces that simply select among more slowly generated, internal genetic variations. The huge population of viruses, combined with their rapid rates of replication and mutation, makes them the world's leading source of genetic innovation: they constantly "invent" new genes. And unique genes of viral origin may travel, finding their way into other organisms and contributing to evolutionary change.

Data published by the International Human Genome Sequencing Consortium indicate that somewhere between 113 and 223 genes present in bacteria and in the human genome are absent in well-studied organisms—such as the yeast *Saccharomyces cerevisiae*, the fruit fly *Drosophila melanogaster* and the nematode *Caenorhabditis elegans*—that lie in between those two evolutionary extremes. Some researchers thought that these organisms, which arose after bacteria but before vertebrates, simply lost the genes in question at some point in their evolutionary history. Others suggested that these genes had been transferred directly to the human lineage by invading bacteria.

My colleague Victor DeFilippis of the Vaccine and Gene Therapy Institute of the Oregon Health and Science University and I suggested a third alternative: viruses may originate genes, then colonize two different lineages—for example, bacteria and vertebrates. A gene apparently bestowed on humanity by bacteria may have been given to both by a virus.

In fact, along with other researchers, Philip Bell of Macquarie University in Sydney, Australia, and I contend that the cell nucleus itself is of viral origin. The advent of the nucleus—which differentiates eukaryotes (organisms whose cells contain a true nucleus), including humans, from prokaryotes, such as bacteria—cannot be satisfactorily explained solely by the gradual adaptation of prokaryotic cells until they became eukaryotic. Rather the nucleus may have evolved from a persisting large DNA virus that made a permanent home within prokaryotes. Some support for this idea comes from sequence data showing that the gene for a DNA polymerase (a DNA-copying enzyme) in the virus called T4, which infects bacteria, is closely related to other DNA polymerase genes in both eukaryotes and the viruses that infect them. Patrick Forterre of the University of Paris-Sud has also analyzed enzymes responsible for DNA replication and has concluded that the genes for such enzymes in eukaryotes probably have a viral origin.

From single-celled organisms to human populations, viruses affect all life on earth, often determining what will survive. But viruses themselves also evolve. New viruses, such as the AIDS-causing HIV-1, may be the only biological entities that researchers can actually witness come into being, providing a real-time example of evolution in action.

Viruses matter to life. They are the constantly changing boundary between the worlds of biology and biochemistry. As we continue to unravel the genomes of more and more organisms, the contributions from this dynamic and ancient gene pool should become apparent. Nobel laureate Salvador Luria mused about the viral influence on evolution in 1959. "May we not feel," he wrote, "that in the virus, in their merging with the cellular genome and reemerging from them, we observe the units and process which, in the course of evolution, have created the successful genetic patterns that underlie all living cells?" Regardless of whether or not we consider viruses to be alive, it is time to acknowledge and study them in their natural context—within the web of life. SA

## Life on the Edge

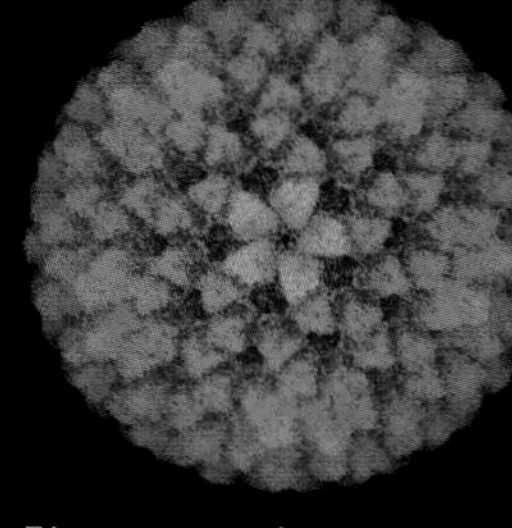
*Bluetongue virus*

**"The very essence of the virus is its fundamental entanglement with the genetic and metabolic machinery of the host."**

*—American Nobel laureate Joshua Lederberg, 1993*

**"Whether or not viruses should be regarded as organisms is a matter of taste."**

*— French Nobel laureate André Lwoff, 1962*

**"A virus is a virus!"** *—Lwoff, 1959*

### MORE TO EXPLORE


**Viral Quasispecies.** Manfred Eigen in *Scientific American*, Vol. 269, No. 1, pages 42–49; July 1993.

**DNA Virus Contribution to Host Evolution.** L. P. Villarreal in *Origin and Evolution of Viruses.* Edited by E. Domingo et al. Academic Press, 1999.

**Lateral Gene Transfer or Viral Colonization?** Victor DeFilippis and Louis Villarreal in *Science*, Vol. 293, page 1048; August 10, 2001.

**Viruses and the Evolution of Life.** Luis Villarreal. ASM Press (in press).

All the Virology on the WWW is at **www.virology.net**

# Are Viruses Alive?

*by Luis P. Villarreal*

# IN REVIEW

## TESTING YOUR COMPREHENSION

1) The word virus comes from a
   - a) Latin word meaning "small particle."
   - b) Latin word meaning "contagion."
   - c) Latin word meaning "poison."
   - d) Greek word meaning "terror."
   - e) Greek word meaning "chance encounter."

2) Wendell Stanley was the first to understand the nature of viruses through his investigation of a(n) _____ virus.
   - a) animal
   - b) plant
   - c) amoebae
   - d) yeast
   - e) bacterial

3) Once a virus gains entry into a cell, its next step is to
   - a) enter the nucleus.
   - b) begin protein synthesis.
   - c) begin synthesis of DNA or RNA, depending on the virus.
   - d) release its genetic material.
   - e) shut down host cell metabolism.

4) A viral genome is
   - a) a viral gene.
   - b) the virus's set of proteins.
   - c) the virus's DNA or RNA.
   - d) the entire virus.
   - e) a partially assembled virus particle.

5) Most viruses
   - a) kill the cell they infect.
   - b) switch between pathogenic and non-pathogenic modes of existence.
   - c) are persistent and innocuous within host cells.
   - d) are derived from bacterial cells.
   - e) are rarely found in the natural world.

6) Viruses can replicate in a cell with a recently destroyed nucleus
   - a) because viral genes are able to direct the cell's metabolic machinery.
   - b) because the virus carries everything it needs without having to depend on the host cell.
   - c) when the virus switches to a bacterial style of replication suitable for cells that lack nuclei.
   - d) when the cell breaks open to allow viral entry.
   - e) only if a key set of cellular genes remain intact.

7) Viruses can play an active role in the evolution of their host species when they
   - a) lose their pathogenic properties.
   - b) evolve drug-resistance.
   - c) acquire the ability to replicate independently of the host.
   - d) become larger and more complex.
   - e) transfer new genes to the host.

8) Two lethally damaged viruses in the same cell may rise from the ashes to create a functioning virus when
   - a) they recombine.
   - b) they acquire new protein coats.
   - c) one of the damaged viruses escapes the cell.
   - d) one of the viruses repairs a critical cellular component.
   - e) one of the viruses takes control of cellular metabolism.

9) The author argues that viewing viruses solely as agents that harm cells prevents us from understanding
   - a) how viruses fit within the web of life.
   - b) the biochemical basis of viral replication.
   - c) how viruses regulate cell metabolism.
   - d) emergent viral infections.
   - e) mechanisms of viral gene mutation.

10) The eukaryotic enzyme that the author proposes is derived from a viral gene is
    a) an RNA synthesizing enzyme.
    b) a DNA copying enzyme.
    c) a DNA repair enzyme.
    d) an enzyme involved in photosynthesis.
    e) an enzyme that recombines DNA.

## BIOLOGY IN SOCIETY

1) Emergent properties are new properties that arise in systems composed of simpler parts. For example, as discussed on page 15, life is an emergent property that arises from collections of molecules that by themselves are non-living. Emergent properties are not restricted to biology. Can you think of any emergent properties of human societies? What individual behaviors underlie these emergent properties? How easy is it to predict emergent properties when the characteristics of system components are known in detail?

2) In experimental human gene therapy, a therapeutic gene is transferred into human cells by incorporating it into the genome of a virus that has been altered to be incapable of replication. Given what you've read about viruses, what kind of dangers might lurk in this approach? Do the potential risks of a virus-based approach to gene therapy outweigh the potential benefits?

3) In 1980, The World Health Organization announced that it had achieved its goal of worldwide eradication of small pox infections. Small pox is a deadly virus that spreads rapidly within a population and is one of the most often discussed bioterrorism agents. There are only two known sites in the world that maintain stocks of small pox virus. However, most scientists believe that clandestine stocks exist that could potentially be controlled by terrorists. Many have argued that the last remaining recognized stocks of virus should be destroyed now that small pox infections have been eliminated. Others counter that these stocks, though potentially dangerous, may provide invaluable research material should terrorists strike by releasing the small pox virus. What should be done with the last recognized stocks of this virus?

4) Viruses are interesting not just for their fascinating biology but also for their profound role in human history. For example, viral diseases largely account for the sharp decline of Native American populations following initial contact with Europeans. Today, the AIDS virus is having a major impact on many populations worldwide. Choose a historical or current event that has been influenced by a virus and speculate how the course of history would have differed if this virus never existed.

## THINKING ABOUT SCIENCE

1) The majority of viruses have a single molecule of DNA or RNA, but some, such as the AIDS and influenza viruses, have more than one. In the case of influenza (the virus that causes flu), seven separate RNA molecules comprise the genome. Focusing on the idea that "viruses are the world's leading source of genetic innovation," what advantage for producing new viral types is offered when viral genomes are composed of separate parts instead of being a single piece of RNA or DNA? You may want to start by thinking of situations in which two or more viruses simultaneously infect a single cell.

2) On page 17, the author reports that more than 100 genes found in both bacteria and humans are absent in eukaryotic model organisms such as yeast, the nematode worm, and the fruit fly. He goes on to list three models to account for this observation. Why is this observation puzzling? How easy it to establish which of the models is correct? What kind of evidence would one need to distinguish between these models? In your answer, present what type of evidence would support each model and how that evidence would make the alternatives unlikely.

3) Some bacteria have evolved to become obligate parasites. This means that the bacterium cannot live on its own outside its host cell. The smallest known bacteria are obligate parasites, some of which have less DNA than large viruses. All viruses are obligate parasites, with the largest virus known, Mimivirus, containing far more DNA than some bacterial parasites. Since some bacteria are obligate parasites and have genomes smaller than the largest viruses, what distinguishes a bacterium from a virus?

## WRITING ABOUT SCIENCE

Are viruses alive? Write an essay in which you answer the question that is the article's title. Include in your essay what it means to be alive, whether there is a distinct line between the living and non-living worlds, and where viruses fall relative to this line or continuum.

**Testing Your Comprehension Answers:**
**1c, 2b, 3d, 4c, 5c, 6a, 7e, 8a, 9a, 10b.**

# THE LITTLEST HUMAN

*A spectacular find in Indonesia reveals that a strikingly different hominid shared the earth with our kind in the not so distant past*

By Kate Wong

**SMALL BUT CLEVER,** ***Homo floresiensis*** **hunts the pygmy** ***Stegodon*** **(an elephant relative) and giant rat that roamed the Floresian rain forest 18,000 years ago.**

On the island of Flores in Indonesia, villagers have long told tales of a diminutive, upright-walking creature with a lopsided gait, a voracious appetite, and soft, murmuring speech.

They call it *ebu gogo*, "the grandmother who eats anything." Scientists' best guess was that macaque monkeys inspired the *ebu gogo* lore. But in October 2004, an alluring alternative came to light. A team of Australian and Indonesian researchers excavating a cave on Flores unveiled the remains of a lilliputian human—one that stood barely a meter tall—whose kind lived as recently as 13,000 years ago.

The announcement electrified the paleoanthropology community. *Homo sapiens* was supposed to have had the planet to itself for the past 25 millennia, free from the company of other humans following the apparent demise of the Neandertals in Europe and *Homo erectus* in Asia. Furthermore, hominids this tiny were known only from fossils of australopithecines (Lucy and the like) that lived nearly three million years ago—long before the emergence of *H. sapiens*. No one would have predicted that our own species had a contemporary as small and primitive-looking as the little Floresian. Neither would anyone have guessed that a creature with a skull the size of a grapefruit might have possessed cognitive capabilities comparable to those of anatomically modern humans.

## Isle of Intrigue

THIS IS NOT THE FIRST TIME Flores has yielded surprises. In 1998 archaeologists led by Michael J. Morwood of the University of New England in Armidale, Australia, reported having discovered crude stone artifacts some 840,000 years old in the Soa Basin of central Flores. Although no human remains turned up with the tools, the implication was that *H. erectus*, the only hominid known to have lived in Southeast Asia during that time, had crossed the deep waters separating Flores from Java. To the team, the find showed *H. erectus* to be a seafarer, which was startling because elsewhere *H. erectus* had left behind little material culture to suggest that it was anywhere near capable of making watercraft. Indeed, the earliest accepted date for boat-building was 40,000 to 60,000 years ago, when modern humans colonized Australia. (The other early fauna on Flores probably got there by swimming

## Overview/*Mini Humans*

- Conventional wisdom holds that *Homo sapiens* has been the sole human species on the earth for the past 25,000 years. Remains discovered on the Indonesian island of Flores have upended that view.
- The bones are said to belong to a dwarf species of *Homo* that lived as recently as 13,000 years ago.
- Although the hominid is as small in body and brain as the earliest humans, it appears to have made sophisticated stone tools, raising questions about the relation between brain size and intelligence.
- The find is controversial, however—some experts wonder whether the discoverers have correctly diagnosed the bones and whether anatomically modern humans might have made those advanced artifacts.

**MODERN INDIAN ELEPHANT**
**(*Elephas maximus*)**

KAZUHIKO SANO (*preceding pages*)

or accidentally drifting over on flotsam. Humans are not strong enough swimmers to have managed that voyage, but skeptics say they may have drifted across on natural rafts.

Hoping to document subsequent chapters of human occupation of the island, Morwood and Radien P. Soejono of the Indonesian Center for Archaeology in Jakarta turned their attention to a large limestone cave called Liang Bua located in western Flores. Indonesian archaeologists had been excavating the cave intermittently since the 1970s, depending on funding availability, but workers had penetrated only the uppermost deposits. Morwood and Soejono set their sights on reaching bedrock and began digging in July 2001. Before long, their team's efforts turned up abundant stone tools and bones of a pygmy version of an extinct elephant relative known as *Stegodon*. But it was not until nearly the end of the third season of fieldwork that diagnostic hominid material in the form of an isolated tooth surfaced. Morwood brought a cast of the tooth back to Armidale to show to his department colleague Peter Brown. "It was clear that while the premolar was broadly humanlike, it wasn't from a modern human," Brown recollects. Seven days later Morwood received word that the Indonesians had recovered a skeleton. The Australians boarded the next plane to Jakarta.

Peculiar though the premolar was, nothing could have prepared them for the skeleton, which apart from the missing arms was largely complete. The pelvis anatomy revealed that the individual was bipedal and probably a female, and the tooth eruption and wear indicated that it was an adult. Yet it was only as tall as a modern three-year-old, and its brain was as small as the smallest australopithecine brain known. There were other primitive traits as well, including the broad pelvis and the long neck of the femur. In other respects, however, the specimen looked familiar. Its small teeth and narrow nose, the overall shape of the braincase and the thickness of the cranial bones all evoked *Homo*.

Brown spent the next three months analyzing the enigmatic skeleton, catalogued as LB1 and affectionately nicknamed the Hobbit by some of the team members, after the tiny beings in J.R.R. Tolkien's *The Lord of the Rings* books. The decision about how to classify it did not come easily. Impressed with the characteristics LB1 shared with early hominids such

PORTIA SLOAN

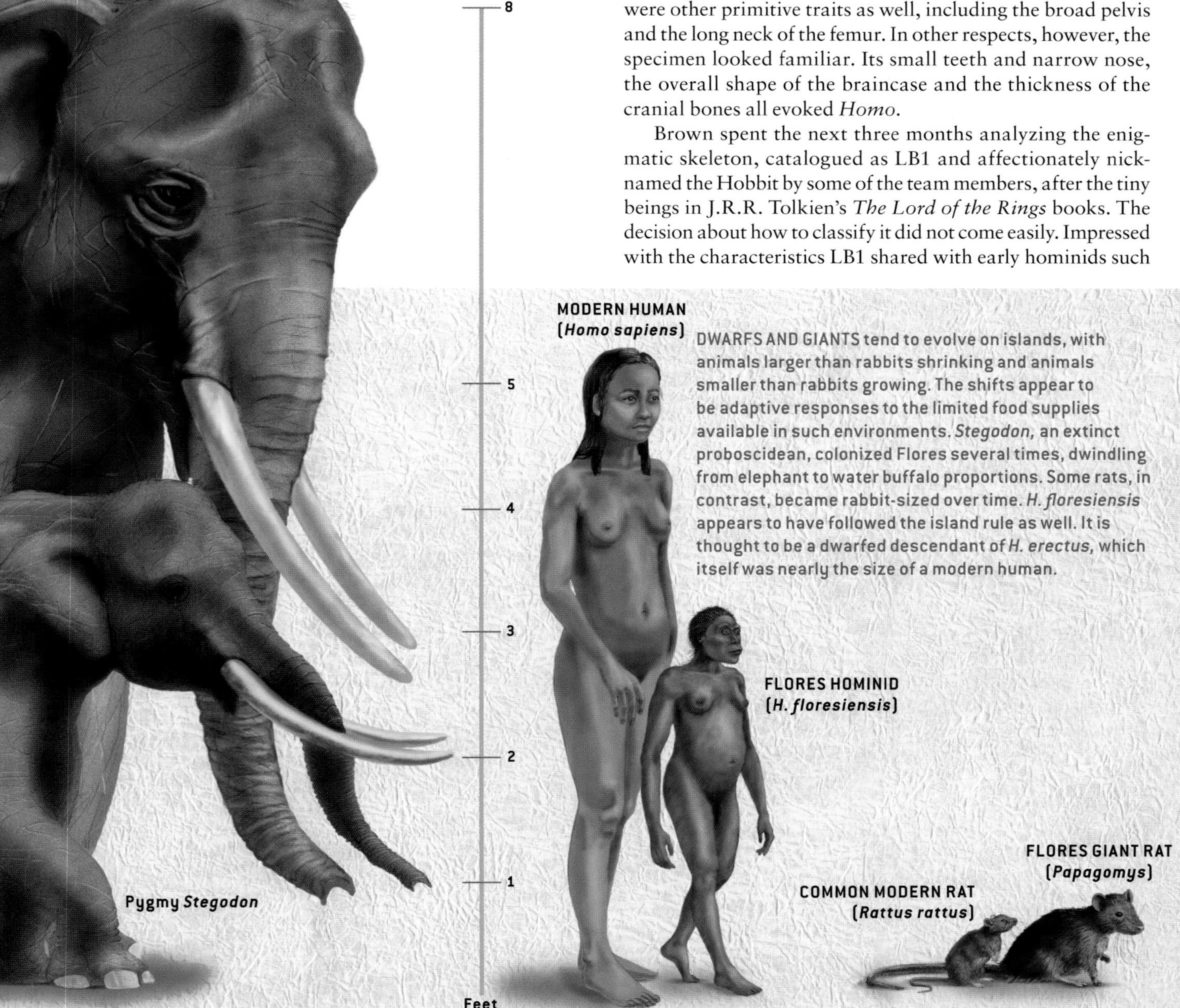


DWARFS AND GIANTS tend to evolve on islands, with animals larger than rabbits shrinking and animals smaller than rabbits growing. The shifts appear to be adaptive responses to the limited food supplies available in such environments. *Stegodon*, an extinct proboscidean, colonized Flores several times, dwindling from elephant to water buffalo proportions. Some rats, in contrast, became rabbit-sized over time. *H. floresiensis* appears to have followed the island rule as well. It is thought to be a dwarfed descendant of *H. erectus*, which itself was nearly the size of a modern human.

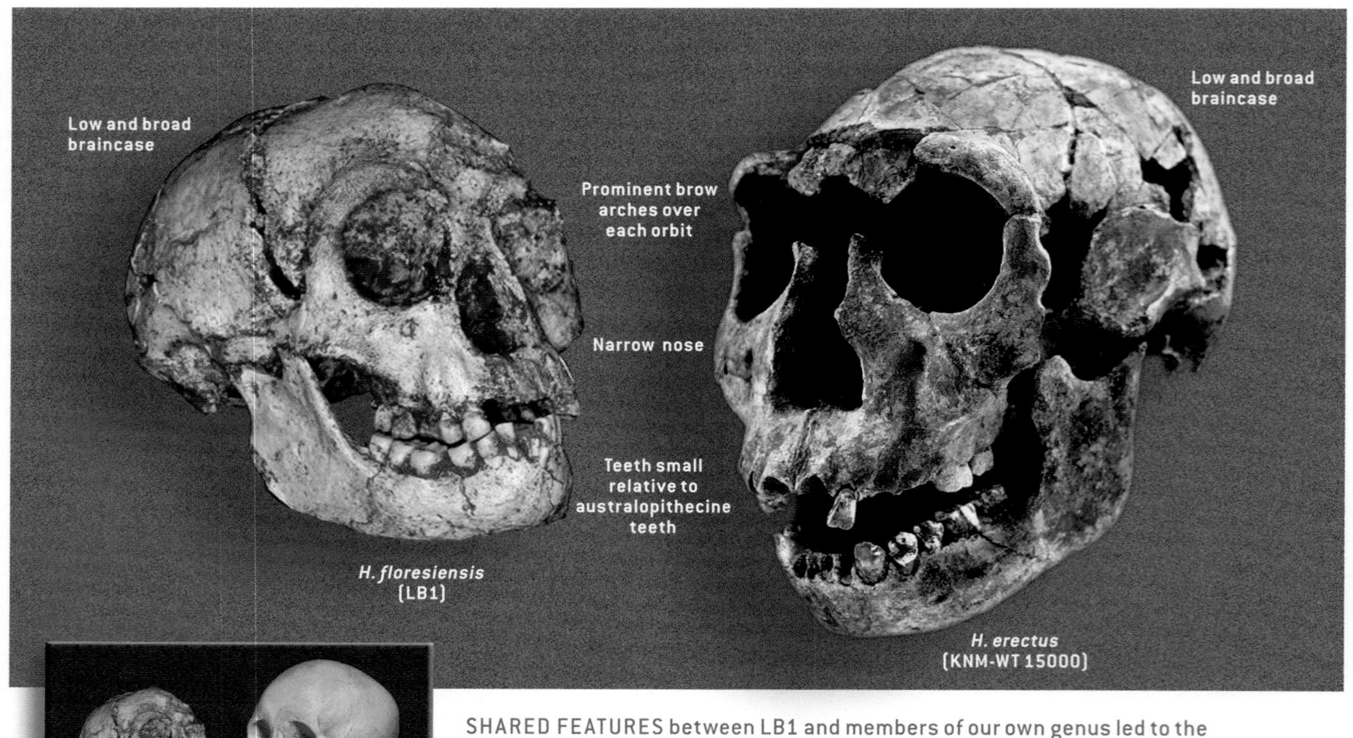


SHARED FEATURES between LB1 and members of our own genus led to the classification of the Flores hominid as *Homo,* despite its tiny brain size. Noting that the specimen most closely resembles *H. erectus,* the researchers posit that it is a new species, *H. floresiensis,* that dwarfed from a *H. erectus* ancestor. *H. floresiensis* differs from *H. sapiens* in having, among other characteristics, no chin, a relatively projecting face, a prominent brow and a low braincase.

as the australopithecines, he initially proposed that it represented a new genus of human. On further consideration, however, the similarities to *Homo* proved more persuasive. Based on the 18,000-year age of LB1, one might have reasonably expected the bones to belong to *H. sapiens,* albeit a very petite representative. But when Brown and his colleagues considered the morphological characteristics of small-bodied modern humans—including normal ones, such as pygmies, and abnormal ones, such as pituitary dwarfs—LB1 did not seem to fit any of those descriptions. Pygmies have small bodies and large brains—the result of delayed growth during puberty, when the brain has already attained its full size. And individuals with genetic disorders that produce short stature and small brains have a range of distinctive features not seen in LB1 and rarely reach adulthood, Brown says. Conversely, he notes, the Flores skeleton exhibits archaic traits that have never been documented for abnormal small-bodied *H. sapiens.*

What LB1 looks like most, the researchers concluded, is a miniature *H. erectus.* Describing the find in the journal *Nature,* they assigned LB1 as well as the isolated tooth and an arm bone from older deposits to a new species of human, *Homo floresiensis.* They further argued that it was a descendant of *H. erectus* that had become marooned on Flores and evolved in isolation into a dwarf species, much as the elephantlike *Stegodon* did.

Biologists have long recognized that mammals larger than rabbits tend to shrink on small islands, presumably as an adaptive response to the limited food supply. They have little to lose by doing so, because these environments harbor few predators. On Flores, the only sizable predators were the Komodo dragon and another, even larger monitor lizard. Animals smaller than rabbits, on the other hand, tend to attain brobdingnagian proportions—perhaps because bigger bodies are more energetically efficient than small ones. Liang Bua has yielded evidence of that as well, in the form of a rat as robust as a rabbit.

But attributing a hominid's bantam size to the so-called island rule was a first. Received paleoanthropological wisdom holds that culture has buffered us humans from many of the selective pressures that mold other creatures—we cope with cold, for example, by building fires and making clothes, rather than evolving a proper pelage. The discovery of a dwarf hominid species indicates that, under the right conditions, humans can in fact respond in the same, predictable way that other large mammals do when the going gets tough. Hints that *Homo* could deal with resource fluxes in this manner came earlier in 2004 from the discovery of a relatively petite *H. erectus* skull from Olorgesailie in Kenya, remarks Richard Potts of the Smithsonian Institution, whose team recovered the bones. "Getting small is one of the things *H. erectus* had in its biological tool kit," he says, and the Flores hominid seems to be an extreme instance of that.

CHRISTIAN SIDOR *New York College of Osteopathic Medicine* (*H. sapiens*); PETER BROWN *University of New England, Armidale* AND TIMEFORKIDS/KRT (*H. floresiensis*); DAVID BRILL (*H. erectus*)

## Curiouser and Curiouser

*H. FLORESIENSIS*'s teeny brain was perplexing. What the hominid reportedly managed to accomplish with such a modest organ was nothing less than astonishing. Big brains are a hallmark of human evolution. In the space of six million to seven million years, our ancestors more than tripled their cranial capacity, from some 360 cubic centimeters in *Sahelanthropus*, the earliest putative hominid, to a whopping 1,350 cubic centimeters on average in modern folks. Archaeological evidence indicates that behavioral complexity increased correspondingly. Experts were thus fairly certain that large brains are a prerequisite for advanced cultural practices. Yet whereas the pea-brained australopithecines left behind only crude stone tools at best (and most seem not to have done any stone working at all), the comparably gray-matter-impoverished *H. floresiensis* is said to have manufactured implements that exhibit a level of sophistication elsewhere associated exclusively with *H. sapiens*.

The bulk of the artifacts from Liang Bua are simple flake tools struck from volcanic rock and chert, no more advanced than the implements made by late australopithecines and early *Homo*. But mixed in among the pygmy *Stegodon* remains excavators found a fancier set of tools, one that included finely worked points, large blades, awls and small blades that may have been hafted for use as spears. To the team, this association suggests that *H. floresiensis* regularly hunted *Stegodon*. Many of the *Stegodon* bones are those of young individuals that one *H. floresiensis* might have been able to bring down alone. But some belonged to adults that weighed up to half a ton, the hunting and transport of which must have been a coordinated group activity—one that probably required language, surmises team member Richard G. ("Bert") Roberts of the University of Wollongong in Australia.

The discovery of charred animal remains in the cave suggests that cooking, too, was part of the cultural repertoire of *H. floresiensis*. That a hominid as cerebrally limited as this one might have had control of fire gives pause. Humans are not thought to have tamed flame until relatively late in our collective cognitive development: the earliest unequivocal evidence of fire use comes from 200,000-year-old hearths in Europe that were the handiwork of the large-brained Neandertals.

If the *H. floresiensis* discoverers are correct in their interpretation, theirs is one of the most important paleoanthropological finds in decades. Not only does it mean that another species of human coexisted with our ancestors just yesterday in geological terms, and that our genus is far more variable than expected, it raises all sorts of questions about brain size and intelligence. Perhaps it should come as no surprise, then, that controversy has accompanied their claims.

## Classification Clash

IT DID NOT TAKE LONG for alternative theories to surface. In a letter that ran in the October 31, 2004 edition of Australia's *Sunday Mail*, just three days after the publication of the *Nature* issue containing the initial reports, paleoanthropologist Maciej Henneberg of the University of Adelaide countered

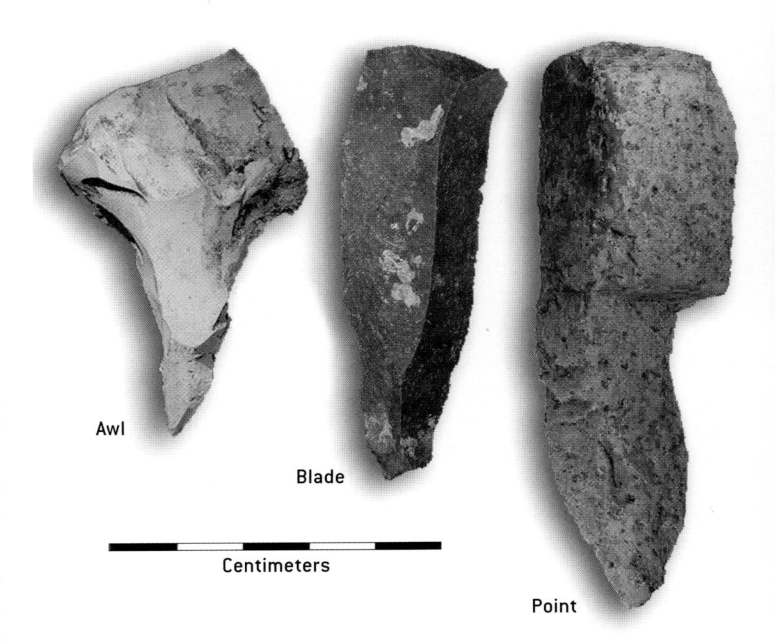


ADVANCED IMPLEMENTS appear to have been the handiwork of *H. floresiensis*. Earlier hominids with brains similar in size to that of *H. floresiensis* made only simple flake tools at most. But in the same stratigraphic levels as the hominid remains at Liang Bua, researchers found a suite of sophisticated artifacts—including awls, blades and points—exhibiting a level of complexity previously thought to be the sole purview of *H. sapiens*.

COURTESY OF MARK MOORE

that a pathological condition known as microcephaly (from the Greek for "small brain") could explain LB1's unusual features. Individuals afflicted with the most severe congenital form of microcephaly, primordial microcephalic dwarfism, die in childhood. But those with milder forms, though mentally retarded, can survive into adulthood. Statistically comparing the head and face dimensions of LB1 with those of a 4,000-year-old skull from Crete that is known to have belonged to a microcephalic, Henneberg found no significant differences between the two. Furthermore, he argued, the isolated forearm bone found deeper in the deposit corresponds to a height of 151 to 162 centimeters—the stature of many modern women and some men, not that of a dwarf—suggesting that larger-bodied people, too, lived at Liang Bua. In Henneberg's view, these findings indicate that LB1 is more likely a microcephalic *H. sapiens* than a new branch of *Homo*.

Susan C. Antón of New York University disagrees with that assessment. "The facial morphology is completely different in microcephalic [modern] humans," and their body size is normal, not small, she says. Antón questions whether LB1 warrants a new species, however. "There's little in the shape that differentiates it from *Homo erectus*," she notes. One can argue that it's a new species, Antón allows, but the difference in shape between LB1 and *Homo erectus* is less striking than that between a Great Dane and a Chihuahua. The possibility exists that the LB1 specimen is a *H. erectus* individual with a pathological growth condition stemming from microcephaly or nutritional deprivation, she observes.

But some specialists say the Flores hominid's anatomy exhibits a more primitive pattern. According to Colin P. Groves of the Australian National University and David W. Cameron of the University of Sydney, the small brain, the long neck of the femur and other characteristics suggest an ancestor along the lines of *Homo habilis*, the earliest member of our genus, rather than the more advanced *H. erectus*. Milford H. Wolpoff of the University of Michigan at Ann Arbor wonders whether the Flores find might even represent an offshoot of *Australopithecus*. If LB1 is a descendant of *H. sapiens* or *H. erectus*, it is hard to imagine how natural selection left her with a brain that's even smaller than expected for her height, Wolpoff says. Granted, if she descended from *Australopithecus*, which had massive jaws and teeth, one has to account for her relatively delicate jaws and dainty dentition. That, however, is a lesser evolutionary conundrum than the one posed by her tiny brain, he asserts. After all, a shift in diet could explain the reduced chewing apparatus, but why would selection downsize intelligence?

Finding an australopithecine that lived outside of Africa—not to mention all the way over in Southeast Asia—18,000 years ago would be a first. Members of this group were thought to have died out in Africa one and a half million years ago, never having left their mother continent. Perhaps, researchers reasoned, hominids needed long, striding limbs, large brains and better technology before they could venture out into the rest of the Old World. But the recent discovery of 1.8 million-year-old *Homo* fossils at a site called Dmanisi in the Republic of Georgia refuted that explanation—the Georgian hominids were primitive and small and utilized tools like those australopithecines had made a million years before. Taking that into consideration, there is no a priori reason why australopithecines (or habilines, for that matter) could not have colonized other continents.

## Troubling Tools

YET IF *AUSTRALOPITHECUS* made it out of Africa and survived on Flores until quite recently, that would raise the question of why no other remains supporting that scenario have turned up in the region. According to Wolpoff, they may have: a handful of poorly studied Indonesian fossils discovered in the 1940s have been variously classified as *Australopithecus*, *Meganthropus* and, most recently, *H. erectus*. In light of the Flores find, he says, those remains deserve reexamination.

Many experts not involved in the discovery back Brown and Morwood's taxonomic decision, however. "Most of the differences [between the Flores hominid and known members of *Homo*], including apparent similarities to australopithecines, are almost certainly related to very small body mass," declares David R. Begun of the University of Toronto. That is, as the Flores people dwarfed from *H. erectus*, some of their anatomy simply converged on that of the likewise little australopithecines. Because LB1 shares some key derived features with *H. erectus* and some with other members of *Homo*, "the most straightforward option is to call it a new species of *Homo*," he remarks. "It's a fair and reasonable interpretation," *H. erectus* expert G. Philip Rightmire of Binghamton University agrees. "That was quite a little experiment in Indonesia."

Even more controversial than the position of the half-pint human on the family tree is the notion that it made those advanced-looking tools. Stanford University paleoanthropologist Richard Klein notes that the artifacts found near LB1

## Home of the Hobbit

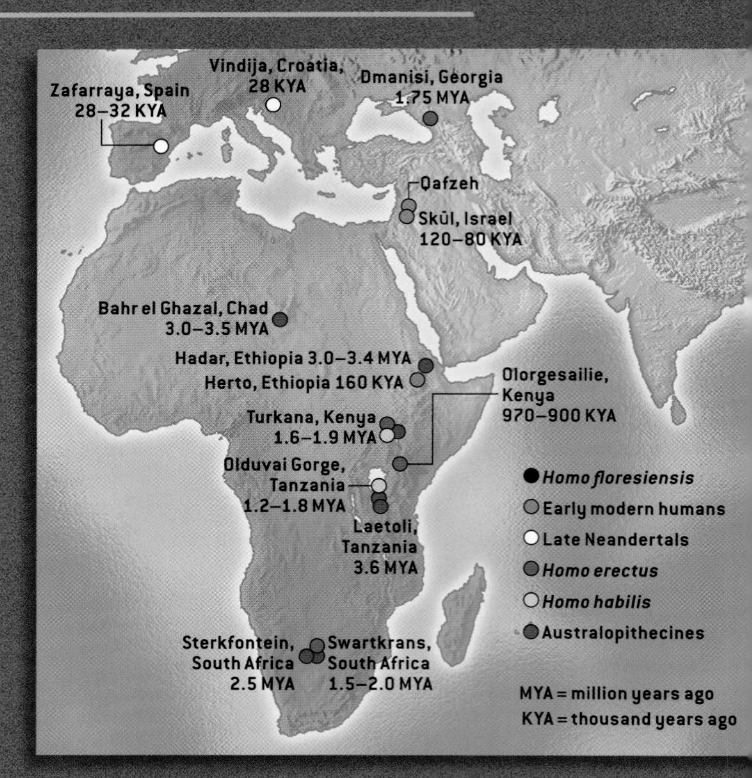


Scholars were stunned a decade ago to learn that *H. erectus* might have survived on the island of Java in Indonesia until 25,000 years ago, well after the arrival of *H. sapiens* in the region and even after the disappearance of Europe's Neandertals. The recent revelation that a third hominid, dubbed *H. floresiensis*, lived in the area until just 13,000 years ago has proved even more provocative.

Archaeologists recovered the remains from a large limestone cave known as Liang Bua located in western Flores. No one knows exactly how humans first reached the island—they may have made the requisite sea crossings by boat, or they may have drifted over on natural rafts quite by accident.

Geographically, Javan *H. erectus* is a good candidate for the ancestor of *H. floresiensis*. But resemblances to specimens from Africa and the Republic of Georgia raise the question of whether *H. floresiensis* stemmed from a different hominid migration into Southeast Asia from the one that gave rise to Javan *H. erectus*. Future excavations on Flores and other Indonesian islands (*detail*) may cast light on these mysteries.

Liang Bua cave

HANDOUT/REUTERS/CORBIS (*cave*); LAURIE GRACE AND EDWARD BELL (*maps*)

appear to include few, if any, of the sophisticated types found elsewhere in the cave. This brings up the possibility that the modern-looking tools were produced by modern humans, who could have occupied the cave at a different time. Further excavations are necessary to determine the stratigraphic relation between the implements and the hominid remains, Klein opines. Such efforts may turn up modern humans like us. The question then, he says, will be whether there were two species at the site or whether modern humans alone occupied Liang Bua—in which case LB1 was simply a modern who experienced a growth anomaly.

Stratigraphic concerns aside, the tools are too advanced and too large to make manufacture by a primitive, diminutive hominid likely, Groves contends. Although the Liang Bua implements allegedly date back as far as 94,000 years ago, which the team argues makes them too early to be the handiwork of

*H. sapiens,* Groves points out that 67,000-year-old tools have turned up in Liujiang, China, and older indications of a modern human presence in the Far East might yet emerge. "*H. sapiens,* once it was out of Africa, didn't take long to spread into eastern Asia," he comments.

"At the moment there isn't enough evidence" to establish that *H. floresiensis* created the advanced tools, concurs Bernard Wood of George Washington University. But as a thought experiment, he says, "let's pretend that they did." In that case, "I don't have a clue about brain size and ability," he confesses. If a hominid with no more gray matter than a chimp has can create a material culture like this one, Wood contemplates, "why did it take people such a bloody long time to make tools" in the first place?

"If *Homo floresiensis* was capable of producing sophisticated tools, we have to say that brain size doesn't add up to much," Rightmire concludes. Of course, humans today exhibit considerable variation in gray matter volume, and great thinkers exist at both ends of the spectrum. French writer Jacques Anatole François Thibault (also known as Anatole France), who won the 1921 Nobel Prize for Literature, had a cranial capacity of only about 1,000 cubic centimeters; England's General Oliver Cromwell had more than twice that. "What that means is that once you get the brain to a certain size, size no longer matters, it's the organization of the brain," Potts states. At some point, he adds, "the internal wiring of the brain may allow competence even if the brain seems small."

LB1's brain is long gone, so how it was wired will remain a mystery. Clues to its organization may reside on the interior of the braincase, however. Paleontologists can sometimes obtain latex molds of the insides of fossil skulls and then create plaster endocasts that reveal the morphology of the organ. Because LB1's bones are too fragile to withstand standard casting procedures, Brown is working on creating a virtual endocast based on CT scans of the skull that he can then use to generate a physical endocast via stereolithography, a rapid-prototyping technology.

"If it's a little miniature version of an adult human brain, I'll be really blown away," says paleoneurologist Dean Falk of the University of Florida. Then again, she muses, what happens if the convolutions look chimplike? Specialists have long wondered whether bigger brains fold differently simply because they are bigger or whether the reorganization reflects selection for increased cognition. "This specimen could conceivably answer that," Falk observes.

## Return to the Lost World

SINCE SUBMITTING their technical papers to *Nature,* the Liang Bua excavators have reportedly recovered the remains of another five or so individuals, all of which fit the *H. floresiensis* profile. None are nearly so complete as LB1, whose long arms turned up during the most recent field season. But they did unearth a second lower jaw that they say is identical in size and shape to LB1's. Such duplicate bones will be critical to their case that they have a population of these tiny humans (as opposed to a bunch of scattered bones from one person). That should in turn dispel concerns that LB1 was a diseased individual.

Additional evidence may come from DNA: hair samples possibly from *H. floresiensis* are undergoing analysis at the University of Oxford, and the hominid teeth and bones may contain viable DNA as well. "Tropical environments are not the best for long-term preservation of DNA, so we're not holding our breath," Roberts remarks, "but there's certainly no harm in looking."

The future of the bones (and any DNA they contain) is uncertain, however. In late November, Teuku Jacob of the Gadjah Mada University in Yogyakarta, Java, who was not involved in the discovery or the analyses, had the delicate specimens transported from their repository at the Indonesian Center for Archaeology to his own laboratory with Soejono's assistance. Jacob, the dean of Indonesian paleoanthropology, thinks LB1 was a microcephalic and allegedly ordered the transfer of it and the new, as yet undescribed finds for examination and safekeeping, despite strong objections from other staff members at the center. At the time this article was going to press, the team was waiting for Jacob to make good on his promise to return the remains to Jakarta by January 1 of 2005, but his reputation for restricting scientific access to fossils has prompted pundits to predict that the bones will never be studied again.

Efforts to piece together the *H. floresiensis* puzzle will proceed, however. For his part, Brown is eager to find the tiny hominid's large-bodied forebears. The possibilities are three-

## The Times of Their Lives

Adding a twig to the family tree of humans, Peter Brown of the University of New England in Armidale, Australia, and his colleagues diagnosed the hominid remains from Flores as a new species of Homo, *H. floresiensis*. This brings the number of hominid forms alive at the time of early *H. sapiens* to four if Neandertals are considered a species separate from our own, as shown here. Brown believes that *H. floresiensis* descended from *H. erectus* (*inset*). Others hypothesize that it is an aberrant *H. sapiens* or *H. erectus* or an offshoot of the earlier and more primitive habilines or australopithecines.

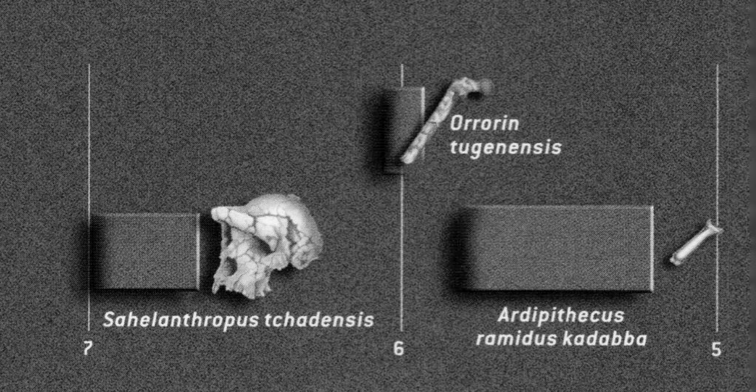


PATRICIA J. WYNNE, CORNELIA BLIK AND EDWARD BELL

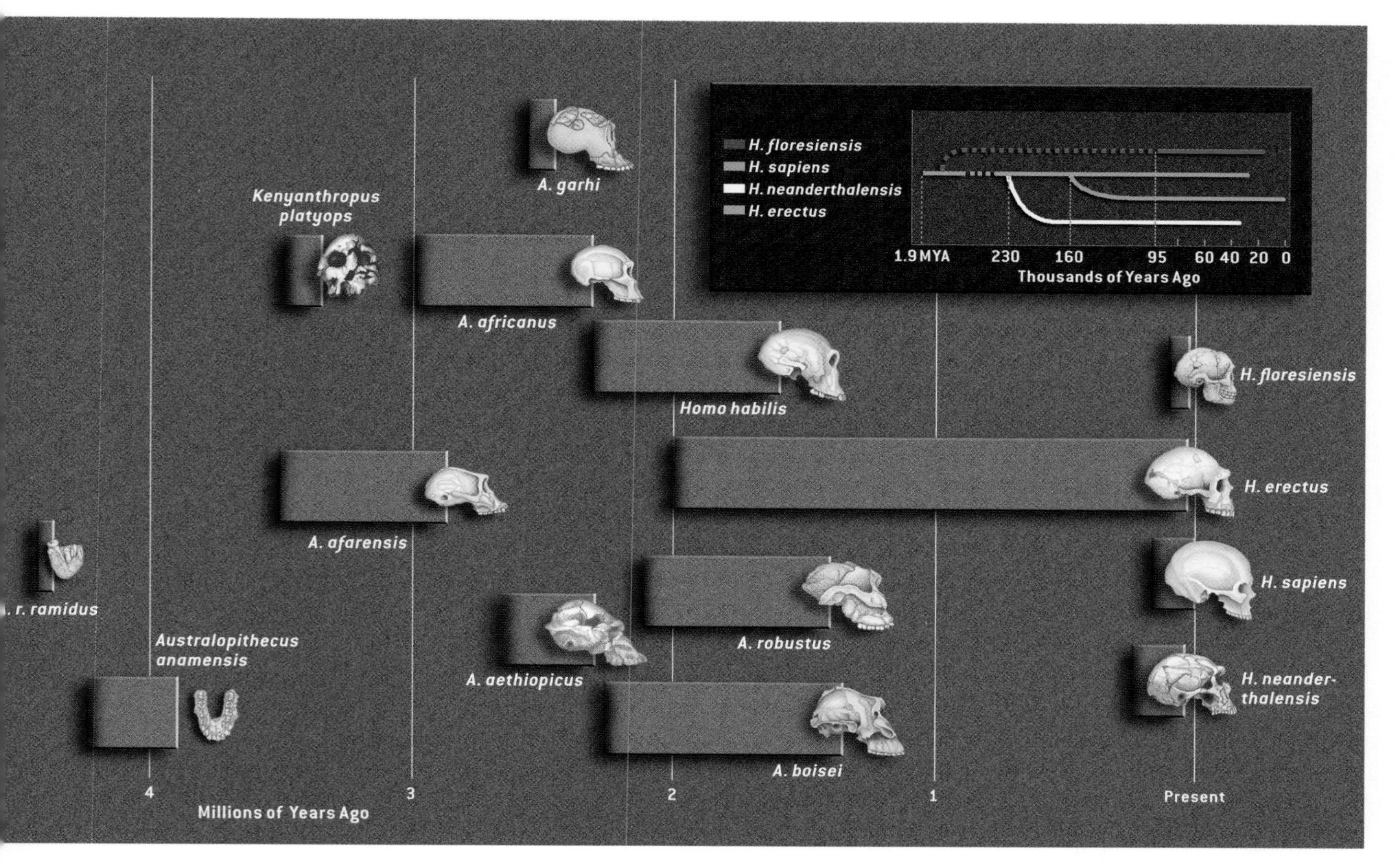


fold, he notes. Either the ancestor dwarfed on Flores (and was possibly the maker of the 840,000-year-old Soa Basin tools), or it dwindled on another island and later reached Flores, or the ancestor was small before it even arrived in Southeast Asia. In fact, in many ways, LB1 more closely resembles African *H. erectus* and the Georgian hominids than the geographically closer Javan *H. erectus,* he observes. But whether these similarities indicate that *H. floresiensis* arose from an earlier *H. erectus* foray into Southeast Asia than the one that produced Javan *H. erectus* or are merely coincidental results of the dwarfing process remains to be determined. Future excavations may connect the dots. The team plans to continue digging on Flores and Java and will next year begin work on other Indonesian islands, including Sulawesi to the north.

The hominid bones from Liang Bua now span the period from 95,000 to 13,000 years ago, suggesting to the team that the little Floresians perished along with the pygmy *Stegodon* because of a massive volcanic eruption in the area around 12,000 years ago, although they may have survived later farther east. If *H. erectus* persisted on nearby Java until 25,000 years ago, as some evidence suggests, and *H. sapiens* had arrived in the region by 40,000 years ago, three human species lived cheek by jowl in Southeast Asia for at least 15,000 years. And the discoverers of *H. floresiensis* predict that more will be found. The islands of Lombok and Sumbawa would have been natural stepping-stones for hominids traveling from Java or mainland Asia to Flores. Those that put down roots on these islands may well have set off on their own evolutionary trajectories.

Perhaps, it has been proposed, some of these offshoots of the *Homo* lineage survived until historic times. Maybe they still live in remote pockets of Southeast Asia's dense rain forests, awaiting (or avoiding) discovery. On Flores, oral histories hold that the *ebu gogo* was still in existence when Dutch colonists settled there in the 19th century. And Malay folklore describes another small, humanlike being known as the *orang pendek* that supposedly dwells on Sumatra to this day.

"Every country seems to have myths about these things," Brown reflects. "We've excavated a lot of sites around the world, and we've never found them. But then [in September 2003] we found LB1." Scientists may never know whether tales of the *ebu gogo* and *orang pendek* do in fact recount actual sightings of other hominid species, but the newfound possibility will no doubt spur efforts to find such creatures for generations to come. SA

---

*Kate Wong is editorial director of ScientificAmerican.com*

## MORE TO EXPLORE


**Archaeology and Age of a New Hominin from Flores in Eastern Indonesia.** M. J. Morwood et al. in *Nature*, Vol. 431, pages 1087–1091; October 28, 2004.

**A New Small-Bodied Hominin from the Late Pleistocene of Flores, Indonesia.** P. Brown et al. in *Nature*, Vol. 431, pages 1055–1061; October 28, 2004.

A Q&A with Peter Brown is at **www.sciam.com/ontheweb**

# The Littlest Human

## *by Kate Wong*

# IN REVIEW

## TESTING YOUR COMPREHENSION

1) The discovery of *H. floresiensis* suggests that our species may have coexisted with another hominid as recently as_____ years ago.
   a) 1,500
   b) 13,000
   c) 25,000
   d) 130,000

2) If humans did colonize Flores as early as 840,000 years ago, this is remarkable because
   a) of the distance the early colonists must have walked from Africa.
   b) Flores was beneath the sea at this time.
   c) there is no evidence for non-hominid animal life on Flores at this time.
   d) it would have required these early people to make a sea crossing.

3) According to the island rule, animals larger than a rabbit evolve to a decreased size because smaller size
   a) allows better heat retention.
   b) requires less nourishment and is not a disadvantage in the absence of large predators.
   c) requires more energy and therefore allows for greater activity.
   d) allows more efficient escape from predators in thick island forests.

4) The discovery of *H. floresiensis* came as scientists were looking for clues about
   a) ebu gogo, the grandmother who eats everything.
   b) possible migration of australopithecines to Flores.
   c) later *H. erectus* habitation of Flores.
   d) the extermination of early hominids by modern humans.

5) A number of scientists argue that the fossil classified as *H. floresiensis* by its discoverers is actually a
   a) *H. sapiens* or *H. erectus* specimen with a growth defect.
   b) Neandertal.
   c) macaque monkey.
   d) very early hominid that lies at the base of the human evolutionary lineage.

6) *H. floresiensis* is most similar to
   a) *H. sapiens*.
   b) *H. neanderthalensis*.
   c) *H. habilis*.
   d) *H. erectus*.

7) The ability of the diminutive *H. floresiensis* to hunt 1,000 pound Stegodons suggests that *H. floresiensis*
   a) possessed language.
   b) could not have been as small as suggested by the single complete fossil.
   c) had tamed fire.
   d) did not consume plant material.

8) The advanced tools found near the *H. floresiensis* fossil
   a) could not have been made by *H. floresiensis*.
   b) were definitely made by *H. floresiensis*.
   c) may have been made by modern humans.
   d) were part of an elaborate hoax.

9) There is _____ correlation between human brain size of normal individuals and their intelligence.
   a) no
   b) a moderate
   c) a strong
   d) a perfect

10) There is evidence that as many as _____ hominids, including *H. sapiens*, co-existed in Southeast Asia for 15,000 years.
a) 2
b) 3
c) 4
d) 5

## BIOLOGY IN SOCIETY

1) Teuku Jacob, the dean of Indonesian paleoanthropology, ordered the *H. floresiensis* remains moved from the Indonesian Center for Archaeology to his own laboratory. One reason for this controversial decision was reported to be Dr. Jacob's desire to have full Indonesian control over fossils discovered in Indonesia, something he believed would not occur at the Center for Archaeology. Who owns fossilized human remains? Who should control access to their study? Should there be an international treaty that mandates ownership and scientific access issues?

2) If *H. floresiensis* followed the island rule and became dwarfed in response to selective pressure, this would contradict a long-standing belief that hominids possessing culture are buffered from selective pressure. How would culture buffer humans from selective pressure? Provide some examples of such buffering. How immune do you think people today are from selective pressure? Why?

## THINKING ABOUT SCIENCE

1) In what ways does the practice of paleoanthropology, a historical science, differ from an experimental science like molecular genetics? In what ways are the practices of a historical and experimental science similar?

2) There is an ongoing debate about whether the remains classified as *H. floresiensis* are a new species in the genus Homo, a new species in the genus Australopithecus, or a small form of *H. erectus*. Each of these views has distinctly different implications for human evolution. How would each view of *H. floresiensis* effect our understanding of hominid evolution? Which of these views is most revolutionary? Which view would make the discovery of *H. floresiensis* least significant?

3) If you were on the team that discovered the *H. floresiensis* remains, what types of evidence would you hope to find to cement your case that *H. floresiensis* was a distinct hominid species that evolved from *H. erectus*? How would the evidence support this view and argue against competing hypotheses?

## WRITING ABOUT SCIENCE

Many pressing economically and medically important problems can benefit from additional research, whereas there is no foreseeable direct economic or heath benefit to studying ancestral hominids. Should precious governmental research funds be committed to paleoanthropological studies such as those of *H. floresiensis*? Write an essay stating your reasons for or against government funding of paleoanthropology. If you argue for funding, what is the appropriate balance between funding for research with clear economic or health benefit and research that has no obvious direct impact on these areas? If you argue against funding, how can you justify a complete lack of support for advancing knowledge of our species' legacy?

**Testing Your Comprehension Answers:**
**1b, 2d, 3b, 4c, 5a, 6d, 7a, 8c, 9a, 10b.**

Wild zebra, asses and horses are being killed for meat, medicine and money. Combined with vanishing habitats and naturally slow reproduction, such predation threatens remaining populations

By Patricia D. Moehlman

# Endangered

AFRICAN WILD ASSES pause on a rocky slope in Eritrea. These young males display the unique pattern of leg stripes that allows researchers to identify individuals.

# Wild Equids

FROM THE TIME OUR ANCESTORS FIRST PAINTED ON CAVE WALLS, the beauty and speed of horses have captured our imagination. During this period, some 20,000 to 25,000 years ago, equids were among the most abundant and ecologically important herbivores on the grasslands of Africa, Asia and the Americas. Today only seven species of wild equids remain—three asses, three zebra and one wild horse—and IUCN-The World Conservation Union now lists most of these as endangered [*see box on opposite page*].

Wildlife biologists, including the Equid Specialist Group of the IUCN, which I chair, study the dwindling populations to learn as much as possible about these historically important animals while they still roam free. We also search for ways to stem their disappearance and have recently developed a plan that prioritizes the actions that should be taken.

## Two Styles of Life

OUR WORK, which builds on that of an early researcher, Hans Klingel of the University of Braunschweig in Germany, distinguishes two distinct patterns of social organization in wild equids. All the animals live in open lands, but their habitats can range from arid desert to grassy plains favored by moderate rainfall. It is the ease of obtaining food and water that determines how these potentially gregarious animals organize themselves for foraging and for mating and rearing their foals.

In the grasslands, such as the Serengeti Plain of Tanzania, abundant forage and water allow females to feed together and thus to form stable groups. A male that can block other males from access to this group gains exclusive mating rights with all the females, and thus this system is referred to as a "harem" or "family." In dry environments, such as the Danakil Desert of Ethiopia and Eritrea, the scattered supply of food and limited water usually do not permit females to forage close to one another or to form consistent groups. Each adult is on its own to find nourishment, and a male will establish a territory near a critical source of water or food; he then controls mating rights with all females that come onto the territory to drink or feed.

In the harem type of organization, groups usually consist of one adult male and one or more females and their offspring. Other males live in "bachelor" groups. The adult females often remain together throughout their lives, but the harem stallion may be displaced by another male, depending on his age and fighting ability and the number of competitors he has to contend with. Foals born into a group stay with it for two to three years before they disperse. Young females usually leave during their first estrus and join other families. Young males tend to stay on for several more years before they depart to find bachelor groups.

The harem strategy, generally followed by plains and mountain zebra as well as by feral horses, often provides a relatively safe environment in which mothers and their foals can thrive. The presence of the dominant stallion markedly reduces harassment from bachelor males, which might otherwise chase and attempt to copulate with the females. Such harassment can be deadly: it hinders the females' ability to feed and can end in abortion or even infanticide. Stable groups and the presence of the stallion also help to fend off predators such as wolves, lions and hyenas.

By contrast, in dry environments, the only long-term assemblages are a female and her offspring, sometimes just a foal, sometimes a foal and a yearling. No permanent bonds persist between adults, although they sometimes form temporary groups. African wild and feral asses, Grevy's zebra and Asiatic wild asses organize themselves in this more socially ephemeral way, with a dominant male controlling a territory near a critical resource. The territorial stallion can dominate his area for years. He tolerates both males and females on his land, but he alone can mate with any female that ventures into his realm.

Controlling access to water is critical. Lactating females need to drink at least once a day, and so they will stay as close to a pond or stream as possible. A female comes into estrus a week or two after giving birth and, if she is not fertilized then, again about a month later. Thus, the territorial male has several

## Overview/*Equid Conservation*

- Wild horses, asses and zebra were once one of the most abundant herbivores in Africa and Asia. Now only seven species remain, and most of these are endangered.
- Human populations, themselves struggling to survive, can be their greatest threat, killing them outright and encroaching on their habitat.
- Extinction is a real possibility for these endangered animals because wild equids reproduce slowly.
- Researchers are stepping up efforts to learn about the animals' way of life and are seeking ways to conserve them in their natural habitats.

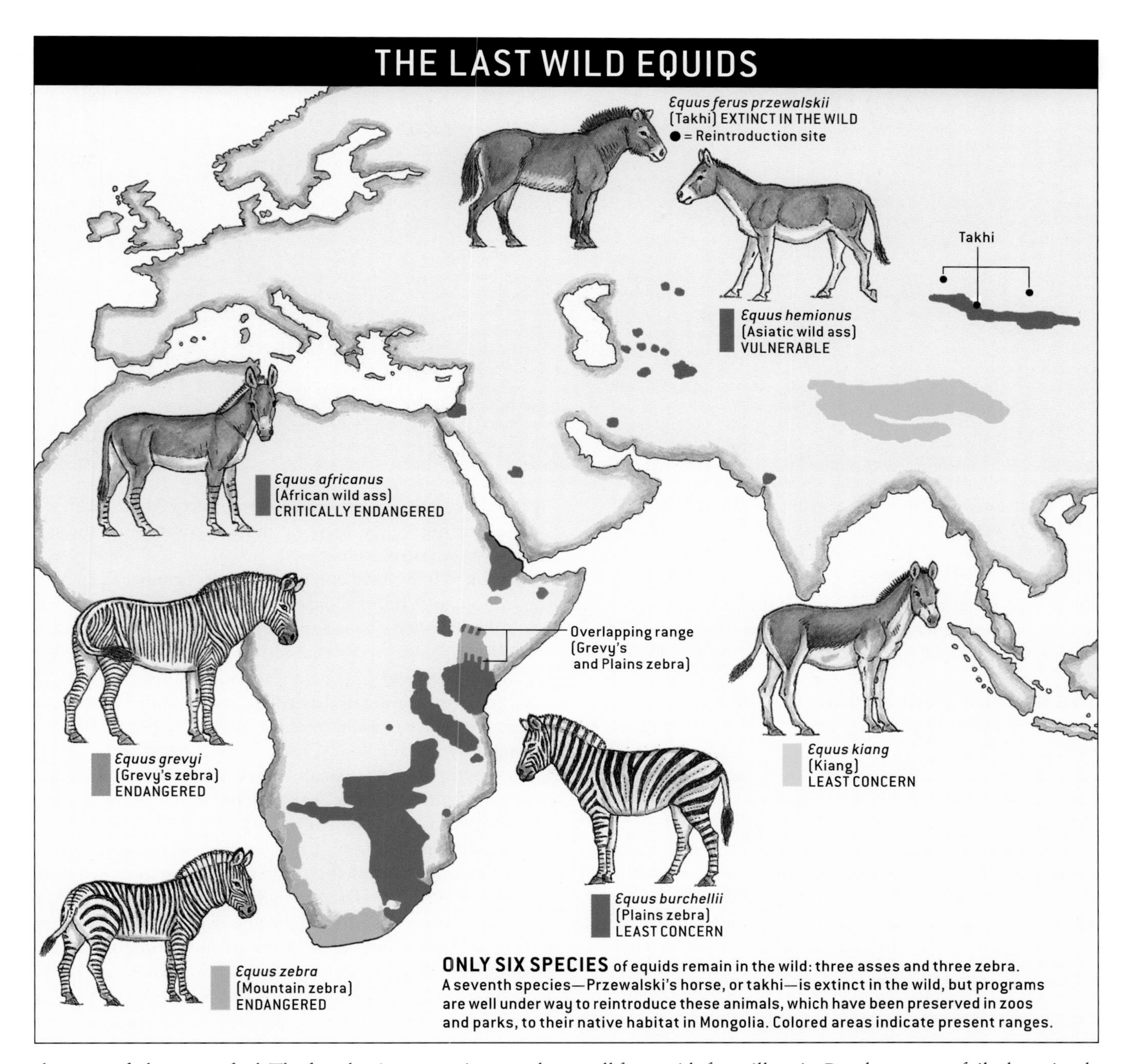


**ONLY SIX SPECIES** of equids remain in the wild: three asses and three zebra. A seventh species—Przewalski's horse, or takhi—is extinct in the wild, but programs are well under way to reintroduce these animals, which have been preserved in zoos and parks, to their native habitat in Mongolia. Colored areas indicate present ranges.

PATRICIA J. WYNNE

chances to father a new foal. The females, in turn, gain not only access to water, they may also benefit from reduced harassment from bachelor males and better protection from predators.

Whichever mating system they follow, the territorial or the harem, all wild equids tend to have their first offspring only after reaching four or five years of age; subsequently, they then reproduce only every other year until the end of their lives at about 16 years of age. Although they have the biological potential to produce a foal every year, they seldom do so in the wild, where the struggle to find food and water restricts reproduction. They nurture their relatively rare offspring with a large investment of parental care—milk, shared food and water, and protection from predators. This kind of trade-off can be a good reproductive strategy, and it worked well for equids for millennia. But the strategy fails the animals when conditions lead to high death rates—such as those that hunters currently levy on equids in their pursuit of food, medicine and the commercial sale of hides.

Death rates are also affected today by loss of habitat and reduced access to forage and water. Females with young foals often have to live farther away from water now, which means that fewer of the foals survive to replenish the population. A small population is more vulnerable than a larger one, because an episode of severe weather or disease can wipe out a geographically isolated group.

Those of us who try to monitor these population trends face a complicated task. Not only do the frequently low numbers of equids in an area make normal sampling techniques

AFRICAN WILD ASS population, which is critically endangered, is beginning to increase in Eritrea because of government support and the conservation ethic of the local Afar people, who share their resources with wildlife.

less effective, but many species live in difficult terrain, which makes finding them a challenge. My own research on the African wild ass (*Equus africanus*) offers a case in point.

## The World's Most Endangered Equid

THE DANAKIL DESERT in the Horn of Africa presents an austere and daunting landscape. Even by desert standards, it is extremely dry; rainfall measures only four inches in a good year. Mountains and ridges of rough lava are furrowed with narrow valleys of alkaline soil sheltering a few grasses and shrubs.

When I set out to search for the African wild ass in the Danakil in 1994, no sightings had been documented for 20 years. Ever since my early research in California during the 1970s on the feral ass in Death Valley, I had been interested in their ancestors in the desert mountains of Africa. At last I was setting out to find them or, more accurately, to find out whether they still existed.

DIFFERENT WILD ASS species, the Kiang, lives on the steppes of the Tibetan Plateau, at a higher altitude than any other equid.

I worked with local conservationists—Fanuel Kebede of the Ethiopian Wildlife Conservation Organization and Hagos Yohannes of the Eritrean Wildlife Conservation Unit. It soon became clear to us that although very few wild asses remained, the local Afar pastoralists knew where we could find these elusive animals. In Eritrea, accompanied by an Afar guide, Omar, we trekked for days, and many hot, dry miles, through the volcanic landscape. Finally, one morning Omar led us up through the basalt ridges of the Messir Plateau. There we found a female, her foal and a male grazing near Afar shepherds tending their sheep and goats.

Since that exciting day, my colleagues and I have identified at least 45 asses that inhabit the plateau. They owe their continued existence and relatively high density in great part to the Afar pastoralists of Eritrea. These people traditionally share their lands and resources with the wildlife and do them no harm. Once they understood the work my colleagues and I were doing, they set out to help. Now when we arrive at their village for a research trip, they round up three camels to carry our camping equipment, food and water, and we all walk to the top of the plateau and set up camp. Thereafter, every other day a man and camel bring us four plastic jerricans with 160 liters of water. This assistance allows us to do our fieldwork on foot in the midst of the best area for the African wild ass.

Just to find this rare and elusive animal ranked as an accomplishment. In the 20 years since wild ass populations were documented in the Danakil, our surveys revealed that their numbers had dropped by more than 90 percent, and the IUCN has designated them as critically endangered; probably fewer than 1,000 (including our 45) remain in the wild. We can tell that the 45 we have located are different individuals, because each animal has a unique pattern of stripes on its legs. Thus, we have been able to follow their movements, social interactions and survival. We can also track a female's reproductive

GREVY'S ZEBRA mother and foal constitute the only stable social unit among these endangered equids that live in the arid habitat of northern Kenya and Ethiopia.

status, how often she gives birth, and the fate of her foals.

What we have uncovered so far tells us that their behavior is typical for equids living in arid habitats: the dominant males maintain mating territories, and the only socially stable group is a mother and her offspring. Occasionally they form small temporary groups made up of fewer than five adults. The composition of these groups varies widely—from single-sex adult groups to mixed groups of males and females of all ages. Females in the same reproductive stage—lactating mothers with foals, for example—may temporarily move and forage together. But competition among females for the sparse forage probably limits their ability to form long-term associations.

Once the male foals reach two to three years of age, we do not see them again in the study area. Presumably they disperse to other areas, suggesting that inbreeding is unlikely. Female foals, in contrast, usually remain with their mother until they produce their own foals.

Our findings about reproductive biology are still limited, but they indicate that females have their first foal at five or six years of age, rather than the more common four or five years, and then may give birth every other year. During prolonged periods of drought, the age at which a female first gives birth may be delayed. Similarly for mature females, a year in which forage is scarce will see few births and few of the foals that do make it into the world will survive. If adult mortality was also high for any reason—because of inadequate nutrition, lack of water or overhunting—the population could decline to such a degree that recovery would be difficult or even impossible.

The years of 1997 and 1998 provided a vivid illustration of how closely reproduction is linked to rainfall. A severe drought on the Messir Plateau in 1997 meant that none of the females had foals. The following year an El Niño brought abundant rainfall to this parched area. All the females had foals, and at least 80 percent of them survived. The potential for such high birth rates and survivorship in good years indicates that the Messir Plateau may be a critical habitat for reproduction. And in fact this area has the highest population density of this species ever recorded—approximately 50 asses per 100 square kilometers. But the highly sporadic rainfall means that the continued existence of the population is precarious.

## A Plan for Survival

IN CONTRAST TO the African wild ass searching for food in their arid habitat, the plains zebra (*E. burchellii*) roam the productive grasslands of Kenya and Tanzania and south to the tip of Africa. They are the most widespread and abundant equid in the world today, although their welfare depends on conservation programs aimed at maintaining their habitat and preventing overhunting. As one would expect, their social organization follows the harem model rather than the territorial. Another species of these striped equids, the Grevy's zebra (*E. grevyi*), lives in a more arid habitat and has the territorial social organization and mating system typical of such land-

THE AUTHOR

*PATRICIA D. MOEHLMAN* received her Ph.D. from the University of Wisconsin–Madison. She has studied the behavioral ecology and the evolution of mating systems in equids and canids for the past 35 years. Since 1989 she has worked with wildlife department personnel and local pastoralists in Somalia, Ethiopia and Eritrea to find and conserve the critically endangered African wild ass. A significant part of her work has involved securing training and postgraduate education for her Ethiopian and Eritrean colleagues. A member of the Wildlife Trust Alliance, she has served as chair of the IUCN-The World Conservation Union/Species Survival Commission Equid Specialist Group since 1997.
Mary Pearl and the Wildlife Trust have provided critical support for the conservation of wild equids. The Whitley Laing Foundation, Saint Louis Zoological Park, Wildlife Conservation Society and African Wildlife Foundation have also provided important funding for the protection of these endangered species.

# The Return of the Takhi

PAINTING of early horse from Lascaux Cave in France.

TAKHI STALLION rounds up mares in his group.

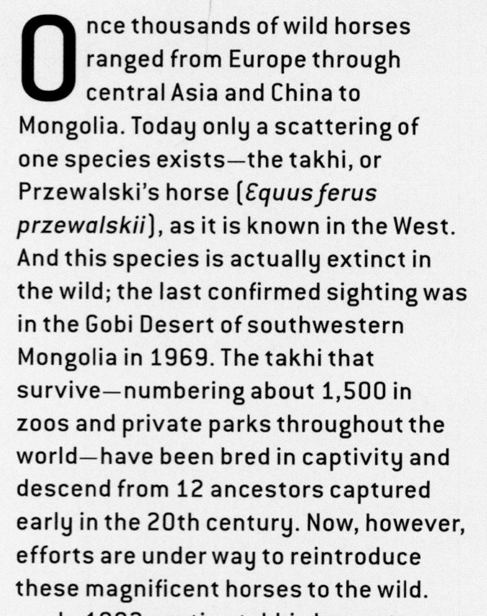

Once thousands of wild horses ranged from Europe through central Asia and China to Mongolia. Today only a scattering of one species exists—the takhi, or Przewalski's horse (*Equus ferus przewalskii*), as it is known in the West. And this species is actually extinct in the wild; the last confirmed sighting was in the Gobi Desert of southwestern Mongolia in 1969. The takhi that survive—numbering about 1,500 in zoos and private parks throughout the world—have been bred in captivity and descend from 12 ancestors captured early in the 20th century. Now, however, efforts are under way to reintroduce these magnificent horses to the wild.

In 1992 captive takhi chosen to represent as much genetic diversity as possible (to avoid the hazards of inbreeding) were flown by transport plane from Europe to two sites in Mongolia: Takhin Tal and Hustain Nuruu. Initially placed in fenced enclosures so that they could adapt to "semiwild" conditions, the horses are now foraging and mating on their native turf. Subsequent transports and births, plus an additional reintroduction site established at Khomin Tal in 2004, have brought the total number of takhi in Mongolia to roughly 250. Since the time of Genghis Khan, the horse has played an integral role the country's culture, and today's Mongolians have welcomed these living symbols of their heritage and have been instrumental in the success of the programs.

Although the takhi is similar to the wild horses that people began to tame some 6,000 years ago, recent DNA research has shown that it is not ancestral to the modern domestic horse. Przewalksi's horse has two more chromosomes than occur in modern domestic horses. The two can interbreed, however, and produce fertile offspring, so the reintroduction programs need to guard against this possibility.

The reintroductions have taught us the critical importance of teaching once confined animals how to avoid predators, such as wolves. And they have alerted us to unexpected problems such as exposure to tick-borne diseases. Even more sobering, we have learned how much it costs to transport and reestablish populations. Saving a species before it goes extinct in the wild would make much better sense. *—P.D.M.*

TAKHI MOTHERS and foals graze at Takhin Tal, Mongolia, one of the sites where these horses have been brought back to their native land. Many foals have been born, but severe winters, exposure to tick-borne diseases, and predatory wolves challenge their survival.

PLAINS ZEBRA live in stable family groups composed of a male and several females with their offspring. The African savannas where they live provide abundant forage, which allows the long-term groups to form.

scapes; these creatures are endangered—only 2,500 to 3,000 remain in northern Kenya and Ethiopia.

Can we then conclude that one system of social organization is more likely to benefit survival than the other? Not necessarily. The Przewalski's horse, or takhi (*E. ferus przewalskii*), shared the harem social system of the plains zebra. Yet these horses are now extinct in the wild [*see box on opposite page*].

Habitat degradation and hunting pressure turn out to present far higher barriers to survival. In its plan for actions to counter these problems, the Equid Specialist Group of the IUCN gives top priority to finding out more about the animals themselves—basic biology, seasonal movements, interactions with livestock, and the dynamics of the arid ecosystems in which they live. Also important are the protection of water supplies, the control of poaching, and improved monitoring of equid populations.

And the Afar pastoralists of Eritrea, with their long-standing practice of sharing resources with wildlife, offer a model for an additional—and essential—component. No attempt to conserve wildlife will succeed without the involvement of the local people. If they have a vital stake in protecting and benefiting from their resources—land, water, vegetation as well as wildlife—then they will have a rationale for investing in the long-term management of this habitat. The income from tourists who come to view the animals in their natural setting may turn out to offer the greatest financial incentive for conserving the environment, but each locale will need to figure out the best strategy for its own constellation of resources and needs. Any revenue from such programs can then be invested in schools, health and veterinary care.

The challenges are formidable, but these steps offer the best chance for the survival of these wonderful animals that have struck awe in the hearts of our own species for thousands of years. SA

## MORE TO EXPLORE


**Horses, Asses, and Zebras in the Wild.** C. P. Groves. R. Curtis Books, Hollywood, Fla., 1974.

**The African Wild Ass (*Equus africanus*): Conservation Status in the Horn of Africa.** P. D. Moehlman, F. Kebede and H. Yohannes in *Applied Animal Behavior Science,* Vol. 60, Nos. 2–3, pages 115–124; November 15, 1998.

**Feral Asses (*Equus africanus*): Intraspecific Variation in Social Organization in Arid and Mesic Habitats.** P. D. Moehlman in *Applied Animal Behavior Science,* Vol. 60, Nos. 2–3, pages 171–195; November 15, 1998.

**Equids: Zebras, Asses and Horses: Status Survey and Conservation Action Plan.** Edited by P. D. Moehlman. IUCN-The World Conservation Union, Gland, Switzerland, 2002.

**Natural and Sexual Selection and the Evolution of Multi-level Societies: Insights from Zebras with Comparisons to Primates.** D. I. Rubenstein and M. Hack in *Sexual Selection in Primates: New and Comparative Perspectives.* Edited by P. M. Kappeler and C. P. van Schaik. Cambridge University Press, 2004.

Equid Specialist Group at the IUCN: **www.iucn.org/themes/ssc/sgs/equid/**

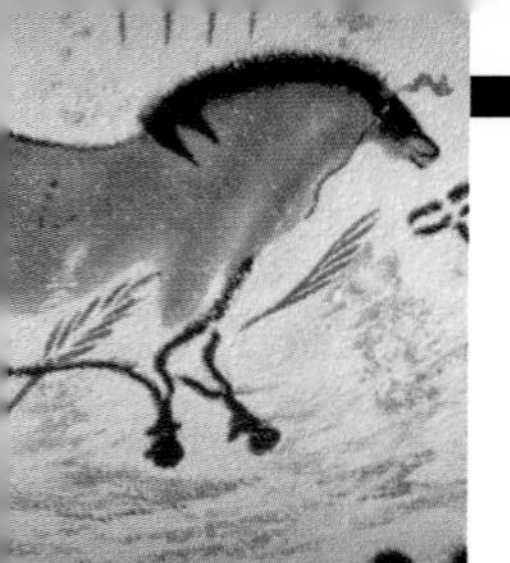

# Endangered Wild Equids

*by Patricia D. Moehlman*

# IN REVIEW

## TESTING YOUR COMPREHENSION

1) An advantage of the harem type of social organization is that it
   a) shortens the gestational period.
   b) increases access to water.
   c) protects females and foals from harassment by bachelor males.
   d) gives more males access to females for mating.
   e) creates stable nuclear families consisting of a male, female and their offspring.

2) There are _____ species of wild equid remaining in the wild today.
   a) 2
   b) 6
   c) 12
   d) 24
   e) 50

3) In the equid territorial social system, dominant males gain access to widely dispersed females
   a) by controlling access to a critical resource.
   b) by a display of elaborate courtship behaviors.
   c) by driving off all other males from the territory.
   d) by seeking out females in estrus as they graze within his territory.
   e) by forming a stable group in which he is the only male.

4) When sexually mature, the female African wild ass gives birth to an average of
   a) 1 foal every 10 years.
   b) 1 foal every other year.
   c) 1 foal each year.
   d) 2 foals each year.
   e) 5 foals each year.

5) The reproductive success of the African wild ass of the Messir Plateau is linked closely to
   a) annual rainfall.
   b) the number of dominant males occupying a territory.
   c) the size of territory controlled by each male.
   d) the strength of hunting pressure from the Afar people.
   e) the severity of tick-born parasitic diseases.

6) The most abundant wild equid is
   a) the plains zebra.
   b) Grevy's zebra.
   c) the Asiatic wild ass.
   d) the African wild ass.
   e) the takhi.

7) Domestication of wild horses began about _____ years ago.
   a) 1,000
   b) 6,000
   c) 15,000
   d) 30,000
   e) 60,000

8) The takhi or Przwalski's horse is
   a) extinct.
   b) a hybrid created by mating of Grevy's and plains zebras.
   c) being reintroduced from captivity into the wild.
   d) one of only two wild equids that are not endangered.
   e) being used in breeding programs with domestic horses to create a robust draft animal.

9) Which system of social organization is most effective for ensuring survival of wild equids?
   a) The harem system.
   b) The territorial system.
   c) Neither of these social systems.
   d) A hybrid between the harem and territorial systems.
   e) Which system is most effective depends on environmental conditions.

10) A key to ensuring the survival of any endangered species is:
   a) enrollment in a captive breeding program.
   b) moving humans off their range.
   c) intensive veterinary care.
   d) involvement of local people in conservation efforts.
   e) manipulating the habitat to improve reproductive success and offspring survival.

## BIOLOGY IN SOCIETY

1) As human populations grow in many developing countries, they exert immense pressure on wildlife in the form of habitat loss and degradation, hunting, and competition for resources. Imagine you are World Bank Program Director for Central Asian Development. The World Bank's mandate is to promote worldwide economic development with an emphasis on less-developed nations. You are particularly interested in sustainable development and want to balance the welfare of nomadic herdsmen with conserving the Asiatic wild ass, a vulnerable species that inhabits grasslands of Central Asia. Populations of Asian wild ass are in decline as more and more grassland is grazed by domestic sheep and goats. Propose a program that can satisfy the economic needs of the herdsmen while preserving their life-style and protecting populations of Asiatic wild ass.

2) An ecosystems service assessment attempts to measure the value to humans of healthy ecosystems. Of the four measures used for ecosystems service assessment, the most difficult to quantify is cultural service, something defined as the value of an ecosystem for recreation, spiritual values, and a sense of place. If you were arguing to the United Nations for strict conservation measures for endangered wild equids, what would you state is the cultural service value of preserving these animals?

3) The author of this article describes her experience with the Afar tribespeople. Initially, the Afar offered little help. Once the nature of author's work became known, a bond was forged between the native people and the investigator. This bond was critical to the success of her research. This story has a happy ending, but unfortunately, not all stories do. What types of situations would make such a cooperative relationship between conservation biologists and local peoples difficult? What solutions might there be to bridging a gap between local interests and those of a conservation biologist?

## THINKING ABOUT SCIENCE

1) Imagine you are reviewing a long-term field study of the social structure of a wild equid. You note that in the early years of the study this species exhibited a territorial type of structure in which the only stable social union was mother-foal pairs. In later years, this species switched to a harem type of social organization. Propose a hypothesis for what was behind the switch between these social structures. What records would you search for to support your hypothesis? If these records did support the hypothesis, could you conclude your hypothesis is correct? If you were fortunate enough to have access to a captive population of these animals in a large reserve, how could you experimentally test your hypothesis?

2) The evolutionary biologist Ernst Mayr stated what we now know as the biological species concept when he defined species as "groups of interbreeding natural populations that are reproductively isolated from other such groups." A potentially significant problem for the reintroduction of the takhi to their ancestral range is that they can interbreed with domestic horses present in this region. The offspring produced from such matings are fertile. Are takhi and domestic horse a single species? If so, how can it be that genetic evidence indicates that takhi are not ancestors of the domestic horse? If they're not the same species, how can they interbreed? Finally, what does the fact that takhi and the domestic horse can interbreed say about limitations in the biological species concept?

3) Small population size, limited genetic variation, or an unstable environment all work against the maintenance of a species. Working alone, any one of these factors is threatening. Working together, these factors conspire to send a population on a downward spiral towards extinction. Propose a model that shows how these factors create a much more dire situation when they work together than when they act alone. To start, you may want to consider why tick-born parasites, which have been around for millennia, are such a concern for small populations of reintroduced takhi. Remember to consider multiple generations, not just one, in your model.

## WRITING ABOUT SCIENCE

There are an estimated 1,000 African wild asses remaining in the wild. Efforts by The World Conservation Union to prevent the extinction of this species are expensive, uncertain of success, and could potentially impede economic development in the Horn of Africa where these animals live. Some argue that since no species remains forever and extinctions are a fact of life, expensive conservation efforts for species too weak to survive on their own are misguided. Write an essay that argues for or against this view, setting out your points for why species conservation programs like the one for the African wild ass are either justified or foolish.

**Testing Your Comprehension Answers:**
**1c, 2b, 3a, 4b, 5a, 6a, 7b, 8c, 9e, 10d.**

Diseases spread by sexual contact have long been among the primary targets of modern medicine. Because of the way they spread, STDs continue to be a major public-health problem—a situation that is unlikely to change

# The Challenges of STDs

By Philip E. Ross

Chronic in the patient, persistent in the population, easy to prevent in principle but not in practice, sexually transmitted diseases (STDs) present some of the greatest challenges faced by applied biological science. STDs were in fact among the first diseases to be attacked by pharmacology and to attract the interest of public-health officials. While modern medicine has reduced the societal impact of many, such as syphilis, particularly in developed countries, they still present serious problems worldwide. And today's most notable STD, human immunodeficiency virus (HIV), the virus that causes AIDS, has become one of the greatest health threats to arise in modern times.

STDs have been a prime target of research from the early days of modern medicine, in large part because they fit so neatly into the foundation of such medicine: the germ theory of disease. They have remained important in part because they generally let their hosts live long enough to create a medical problem, a marketing opportunity and even a political lobby. Yet rational treatment of the diseases—and, still more, their rational prevention—have been complicated by nonmedical issues. More than any other type of disease, STDs have been associated in the popular mind with moral weakness and social decadence.

This association creates a potential conflict of interest between patients and their physicians, on the one hand, and public-health authorities, on the other. The doctor-patient relationship has been considered sacrosanct since the days of Hippocrates. Yet in the case of STDs, this respect for privacy and confidentiality must sometimes yield to wider concerns. Municipalities regularly quarantine people with dangerous, contagious illnesses, such as drug-resistant tuberculosis, but these quarantines are accompanied by special medical treatment, and can thus be portrayed as for the good of the patient as well as the community. In contrast, tracing a patient's prior sexual contacts often goes directly against his personal interests, given the embarrassment and shame often involved. It is therefore exceedingly difficult for doctors and public-health officials to get forthright and honest answers from patients.

Since not all cases of STDs are even recognized by those affected, let alone referred to doctors and reported to public-health officials, it can be hard to get a full picture of the health burden they impose. The Centers for Disease Control (CDC) recently attempted to come up with a comprehensive estimate, pegged to the year 1998. It found that in that year, STDs accounted for some 20 million "adverse health consequences" in the U.S., including both the obvious (doctors' time, hospitaliza-

## COMMON STDs

| DISEASE | PATHOGEN TYPE | PATHOGEN NAME | EARLY SYMPTOMS | COMPLICATIONS | INCIDENCE IN THE U.S. | TREATMENT | PREVENTION |
|---|---|---|---|---|---|---|---|
| *Genital herpes* | Virus | *Herpes simplex virus (HSV)* | Cold sores on genitals or mouth | Recurrent outbreaks | 20 percent exposed | Uncurable; symptoms treatable by antivirals | Condoms, use of antivirals |
| *Chlamydia* | Bacterium | *Chlamydia trachomatis* | Burning during urination; often asymptomatic, especially in women | Infertility | 3 million new cases per year | Curable with antibiotics | Condoms; routine screening |
| *AIDS* | Retrovirus | *Human immuno-deficiency virus (HIV)* | Flulike fever | Opportunistic infections | 40,000 new cases per year | Antiretroviral combination therapy | Condoms |
| *HPV* | Virus | *Human papilloma virus (HPV)* | Genital warts; often asymptomatic | Cervical cancer | 80 percent of women by age 50 | Uncurable; warts and precancerous lesions removable by surgery | Condoms; routine screening; Merck is preparing for market a vaccine against HPV, the first STD vaccine ever. |
| *Trichomoniasis* | Protozoan | *Trichomonas vaginalis* | In women, yellow-green vaginal discharge; often asymptomatic in men | Secondary infections; premature birth | 7 million new cases per year | Curable with metronidazole | Condoms; curing male sex partners of infected women |
| *Gonorrhea* | Bacterium | *Neisseria gonorrhoeae* | In men, burning during urination; discharge, in women often asymptomatic | Infertility, blindness in newborns | 700,000 new infections per year | Curable with antibiotics | Condoms |
| *Molluscum contagiosum* | Virus | *Molluscum contagiosum virus (MCV)* | Lesions on thighs, buttocks, lower abdomen; can last for years | Can progress in immuno-compromised patients | Rare; <3 percent of STDs | May resolve; lesions can be removed surgically or chemically | Little protection from condoms; avoid skin-to-lesion contact |
| *Chancroid* | Bacterium | *Haemophilus ducreyi* | Soft, pus-filled sore, often painful in men; often undetect-able in women | Lymph-gland swelling, secondary infection | Very rare, but rising | Curable with antibiotics | Condoms; avoid touching sores |
| *Human t-Cell lymphotropic virus* | retrovirus | HTLV-I and HTLV-II | Generally none | Sometimes causes leukemia | Rare | Early detection to improve chances of curing leukemia | Condoms |
| *Ectoparasites* | lice | *Pthirus pubis* | Itching in genital region | No serious complications | 3 million new cases per year | Curable with medicinal creams and shampoos | Avoid contact with infested people, bedding, clothing |
| *Syphilis* | bacterium | *Treponema pallidum* | Hard, rubbery chancre on genitals or skin | Blindness, mental illness, heart disease | 7,000 new cases per year | Curable with antibiotics | Condoms |

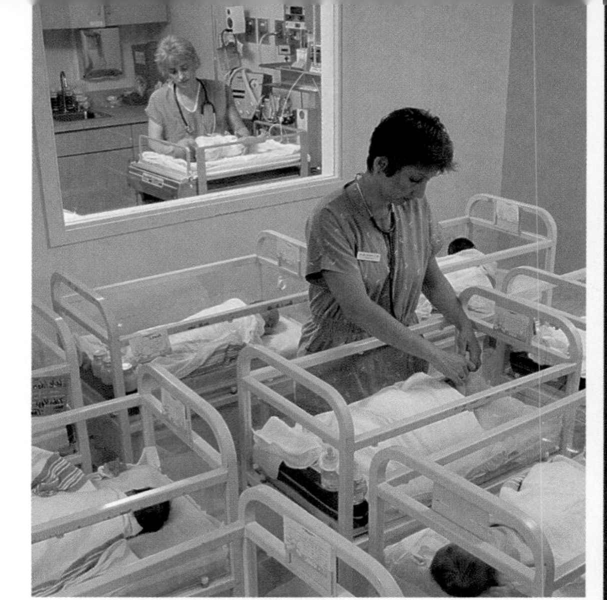

MODES OF TRANSMISSION: Unprotected rivers, such as the Yamuna in India (right) have been infamous disseminators of disease. Pathogens carried by a vector, such as a mosquito (bottom left), need not spare the host: vectors can transmit the pathogen regardless of the host's state. Hospital attendants, who may touch many babies (top left), can also inadvertently serve as vectors.

JIM PICKERELL, *TONY STONE WORLDWIDE (top left)*; WILLIAM C. BROWN, *SCIENCE SOURCE/PHOTO RESEARCHERS INC. (top right)*; JOSEPH POLLEROSS/ JB PICTURES *(bottom left)*

tion) and the harder-to-gauge (lost work hours, infertility treatments). Almost half the burden fell on women, a figure that rises to nearly 90 percent if one removes AIDS from the equation. STDs also resulted in nearly 30,000 fatalities that year—enough to make them one of the country's top 10 causes of death.

Eliminating STDs, or at least reducing their prevalence, is therefore a worthy goal of the public-health community. To do so, scientists must first understand why these diseases behave the way they do. For although many STDs (particularly HIV) can be lethal to their hosts, they are most dangerous to society at large when in hosts who are still alive.

## The Evolutionary Biology of STDs

There are many vehicles that allow disease-causing germs, or pathogens, to travel from one person to the next. Some of the most common are droplets in the air (measles, for instance); feces in untreated water (cholera); the transfer of mothers' blood to their newborn infants, in what is known as congenital transmission (syphilis, HIV); and intermediary hosts, called vectors—which can be anything from mosquitoes (malaria) to hypodermic needles (hepatitis, HIV). STDs use some of these pathways, but their main vehicle, of course, is sexual contact.

For any pathogen, each route poses trade-offs of its own. For instance, droplet-borne bugs have a short range but can spread very quickly. Mosquito-borne bugs tend to spread more slowly, but they do not require that their hosts be crowded together. Yet no matter how a pathogen spreads, the details of its environment make a big difference. If those details change, some strains can benefit and proliferate like mad, while others, less suited to the new conditions, can quickly fade away. In recent years, biologists who attempt to understand infectious diseases in these terms have created the discipline of evolutionary medicine. As an applied science, it aims to prevent the spread of pathogens by exploiting their weaknesses.

Consider the case of a pathogen that passes directly from one person to another via airborne droplets. A strain that reproduces quickly inside a host will have the advantage of inducing violent sneezes and coughs, making the sick person highly infectious. Such a strain will spread fast so long as its host comes into contact with fresh hosts, as in the crowded environment of a refugee camp, troop ship or hospital ward. But if patients are ill enough to retreat to their beds, far from other people, then the strains that make people this sick will die out, and forms that make their hosts less sick will win the race for survival. This is why most pathogens that spread on airborne droplets, such as the common cold, cause only minor sniffles and sneezes.

Now consider the malaria parasite, a one-celled animal that reaches a new host by hitching a ride on a mosquito. This

**More than any other type of disease, STDs have been associated in the popular mind with moral weakness and social decadence.**

parasite can still succeed if it multiplies as furiously as it can, because it can be transmitted even if the host is so feverish that he has to stay in bed. The parasite will still be able to jump to new hosts for a simple reason: mosquitoes pay house calls. Illnesses that spread through such vectors thus tend to make their patients very sick indeed.

The virulence of STDs similarly depends on the underlying details. In a world in which people very rarely come into sexual contact with new partners, no STD pathogen can propagate quickly. For a given strain to survive, therefore, its hosts must not only survive for a long time, but they must both feel well enough to want sex and remain attractive enough to be able to find partners.

Imagine, now, that the world changes, and people begin to come into contact with a great many sexual partners. Populations that had been isolated sexually come to form links, perhaps in the form of traveling salesmen, tourists and others who often cross social borders. With such changes, faster-reproducing STD strains—even though they may kill their hosts very quickly—will predominate, because they can still spread rapidly to new hosts. These diseases will tend to make people sicker, faster. It is therefore no accident that some very serious STDs, notably HIV, arose when such economic and cultural changes took place worldwide.

Educating the public to avoid the more dangerous sexual practices—having vaginal or anal sex without a condom, or with multiple partners whose health status is unknown—therefore pays off in two ways. First, the policy is good medicine, because it lowers the chance of any given person getting the disease. Second, it is good public-health practice, because it should push STDs to evolve into milder forms.

Because their current hosts must remain alive for a while to allow disease propagation, most STDs have evolved into forms that do not reproduce too quickly. One might think that this constraint would keep the concentration of pathogens in the blood down, making hosts somewhat non-infectious. Yet people with STDs tend to be quite infectious. There are several reasons why this is so. Unlike most pathogens, those that cause STDs tend to live out their lives in a permanent envelope of bodily fluids that protects them from drying and other mishaps. Also, STD pathogens are injected into parts of the body that are only lightly policed by the body's immune system. What is more, as noted above, STDs sometimes spread by means other than sexual contact.

One other general advantage that STDs have over most other infectious diseases is their ability to re-infect people again and again. For reasons that are still not entirely understood, people do not naturally become immune to these microorganisms. Scientists have therefore had great difficulty in devising effective vaccines against them.

The different aspects of the public-health problems posed by STDs are best illustrated by specific examples. Let us look, first, at syphilis, the disease that sparked the development of the medical and public-health methodology that shapes policy options to this day, before moving on to examine some of the other most common STDs: chlamydia, herpes and HIV.

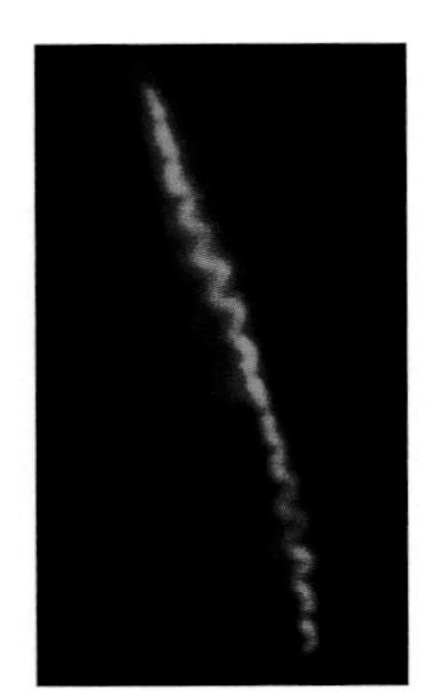

## Syphilis

The mother of all STDs, syphilis was first noted by physicians in Europe in the late 15th century and hence blamed, by many, on Columbus's returning sailors, who were thought to have brought it back from the New World. Some scientists now argue, instead, that the agent of syphilis, a spiral-shaped bacterium called *Treponema pallidum*, had long caused related diseases that were transmitted by skin-to-skin contact (common when people huddled for warmth under shared blankets). If they are right, it may be

speculated that something in the environment changed, favoring the success of those strains of *T. pallidum* that were particularly suited to spreading through sexual contact. (The origin of syphilis remains a controversial topic among anthropologists.)

When it first became an epidemic disease, syphilis progressed so quickly—sometimes killing a patient within months—that it was easy for physicians to deduce its sexual mode of transmission. Later, syphilis became milder, in the sense that the disease would persist in a chronic state over many years.

There is about a 30 percent chance of contracting syphilis from a single exposure during unprotected sex, although the risk varies depending on the stage of the infection, the nature of the sexual practices and the strength of the immune system. The first symptom comes several weeks after exposure, when a hard or rubbery sore, called a chancre, develops at the bacterium's point of entry into the body. Untreated, the chancre will heal within a few weeks; then the disease will enter a new phase, called secondary syphilis. Here the symptoms vary enormously but often include fatigue, rash and a sore throat. Because these symptoms often mimic those of other illnesses (hence the disease's nickname, "the Great Imitator"), diagnosis can be difficult.

Then the symptoms die away, and the patient has no clue that he is still sick, although he remains infectious. This so-called latent period can last just a few months, but in more than half of all cases, it goes on for the rest of the life of the patient. Those who do progress to the tertiary phase of the disease develop painful tumors, bone degeneration, blindness, cardiovascular problems and neurological disorders, which can lead to personality changes, dementia and death.

Syphilis was originally treated crudely, with mercury and other substances that proved nearly as toxic to the patient as to the pathogen. Early in the 20th century, German microbiologist Paul Ehrlich began a quest for what he termed the "magic bullet"—a drug that would kill the syphilis pathogen but spare the host. A year later, after screening hundreds of compounds, Ehrlich's colleague Sahachiro Hata discovered one containing arsenic that at least approached their ideal. The drug, called arsphenamine and trademarked Salvarsan, may be said to have spawned the modern, research-oriented pharmaceutical industry.

A later refinement of the arsphenamine regimen, incorporating other medicines, only had variable effects and took at least 60 weeks for patients to follow its full course. It was only in the 1940s, with the introduction of penicillin, that a certain cure was demonstrated. Even then, early penicillin regimens required repeated injections. Patients often failed to finish the course of treatment, which is terrible from the public-health viewpoint for two reasons: the patient can go on to infect others, and the surviving pathogen is given the opportunity to evolve resistance to the drug. Nowadays, versions of penicillin and other antibiotics are available in slow-release doses that need be administered only once to eliminate the infection.

In the end, then, with penicillin, the medical problem of syphilis could be considered solved. But in order to eliminate the disease utterly, it was also necessary to find and cure all its carriers. This goal required that public-health officials trace the sexual contacts of every known patient, a particularly intrusive practice. People tend to be secretive about their sexual relations, and are often reluctant to let sexual partners know that they have given them a notorious disease. Laws, therefore, were passed in the first half of the 20th century requiring physicians to report syphilitic patients to public-health authorities, who would then conduct contact-tracing assiduously and urge those who had been exposed to undergo treatment.

These methods have succeeded in nearly eliminating syphilis from entire countries for years at a time. In the U.S., the incidence rose in the 1980s and early 1990s, then fell considerably, dropping to 32,000 new cases in 2002, according to the CDC. (Worldwide, however, the disease remains a scourge: the World Health Organization estimates there are about 12 million new cases worldwide annually.) The number of syphilis infections serves as an indirect measure of the effectiveness of a public-health system, as well as of the prevalence of safe sexual practices. When syphilis gets out of hand, it is often a sign that other, perhaps harder-to-track STDs may also be spreading.

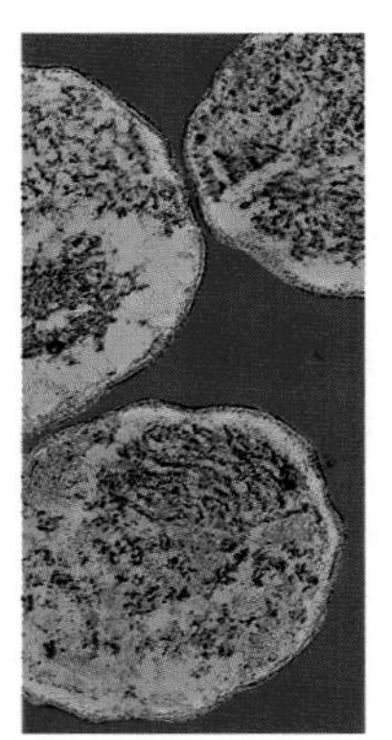

## Chlamydia

The public-health dimension of STDs is well illustrated by chlamydia, a rather common family of bacteria whose sexually transmitted form—*Chlamydia trachomatis*—often produces no symptoms, particularly in women. It thus constitutes a "silent" disease that is particularly hard to control. Chlamydia is the most common STD in the U.S., with some 830,000 reported cases a year, and an estimated three million cases in total. Worldwide, there are more than 90 million new cases a year.

In men the bacteria generally infect the urethra, sometimes causing a burning feeling during urination and producing a discharge, sometimes not. Long-term complications in male patients are rare. Women notice such symptoms in only about 15 percent of cases, but they are much more likely to suffer from complications. If the infection spreads to the fallopian tubes—which carry eggs from the ovaries to the uterus—it may provoke the body's immune system to mount a vigorous defense, in the form of widespread inflammation. It is this inflammation and the scarring it leaves behind, rather than the bacteria themselves, that block the tubes, causing infertility. Often a woman never knows she has had a chlamydia infection until she tries to get pregnant and fails. Both men and women who engage in anal sex can also get infections in the rectum, leading to pain and bleeding or some other discharge.

Treatment is easy; a single dose of the antibiotic drug azithromycin will clear up the infection. The hard part is

KARI LOUNATMAA, *SPL/PHOTO RESEARCHERS, INC.*

diagnosing the ailment in time to head off complications and warn sexual partners to seek treatment. Sexually active people, particularly women, can be screened via samples of their urine or genital secretions. Yet even such screenings are foolproof only if conducted quite frequently.

Methods to prevent chlamydia infection in the first place are under study now. Some researchers are investigating gels and foam that could be applied to shield target tissues from infection. Others are working on vaccines. Still others are trying to find drugs that would slow the bacteria's propagation within the body and ease the task of the body's immune system. For now, though, the only proven method of fighting the disease is by sex education, in which half the job is simply motivating women to undergo screening.

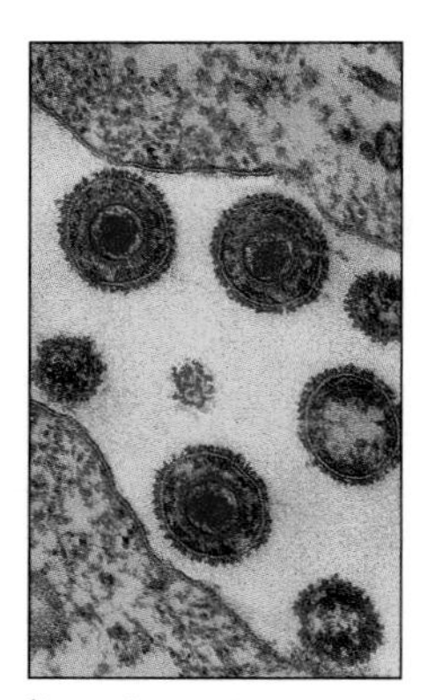

## Herpes

Genital herpes, though it rarely involves complications and can be effectively treated, if not cured, was nonetheless among the most dreaded STDs in the country at one time. In 1982 it was featured on the cover of *Time* magazine, which predicted that it would cause sweeping social changes, notably a return to chastity. Although the disease is in fact quite common—the CDC estimates that about 45 million Americans, or one in five adolescents and adults, have been exposed to it—neither the dread nor the predictions have proved justified.

The STD is caused by a strain of the herpes simplex virus, the cause of cold sores around the mouth. Indeed, the genital form is only a slight variation on the oral one, from which it appears to have evolved, perhaps under the influence of changing sexual practices. As with oral herpes, the first symptoms come some three to 10 days after exposure, when tiny sores break out near the virus's site of entry. The sores break, form a scab and heal without leaving a scar, all in a few weeks. Then the virus holes up in nerve fibers, becoming dormant for months or even years before staging new outbreaks. These "secondary" outbreaks are generally milder and shorter than the initial one, and patients can often feel them coming a day or two in advance. Such premonitions make it easier to minimize the symptoms with antiviral medications, such as acyclovir.

It is easier for a man to pass the virus to a woman than the reverse. One reason may be that the female genitalia are more likely to suffer tiny wounds during sex that facilitate the virus's entry. Condoms thus help to prevent the spread of herpes. The best preventative is to abstain from sex during outbreaks and to quell symptoms (and infectiousness) with antiviral medications. But *Time's* prediction notwithstanding, genital herpes poses a serious problem only to newborns—in whom it can spread to the brain and internal organs—and to adults whose immune systems have been compromised.

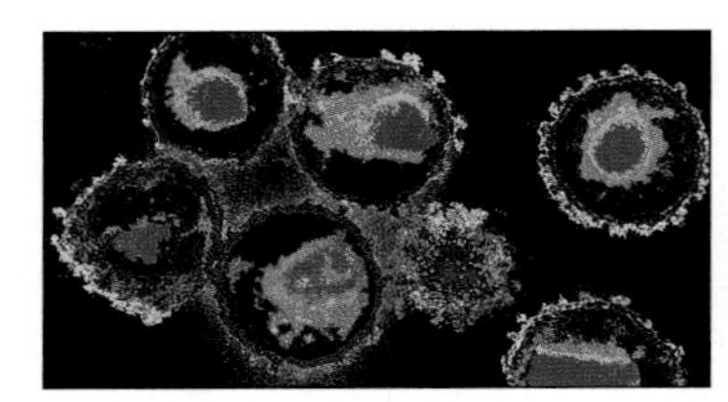

## HIV

When, in 1984, AIDS was traced to a single causative agent, HIV, no one expected that the quest for a solution would be so arduous. Margaret Heckler, then secretary of the U.S. Department of Health and Human Services, famously predicted that a protective vaccine would come within a matter of years. In fact, immense progress has been made in understanding the evolution of the virus and in devising chemotherapies against it, and as a result, those who have access to modern medical treatment are now living much longer with the infection than was the case at first.

HIV reaches the bloodstream from its point of entry (the genitalia, rectum or elsewhere) and, in its initial stage, produces flulike symptoms. It takes refuge in the lymph nodes, a critical part of the immune system, where it attacks what are known as CD4+ T cells (sometimes called helper T cells), which serve as quarterbacks to coordinate the actions of other immune cells. Normally CD4+ T cells are present at a level of about 1,000 cells per cubic milliliter of blood; when this concentration drops below about 200 per cubic milliliter, the body begins to lose its ability to combat a range of normally harmless viruses and microbes. These agents then stage opportunistic infections.

Some early observers understandably mistook these opportunists for the ultimate cause of AIDS, especially since it is these infections, rather than HIV itself, that normally sicken and kill the patient (although HIV alone can do so, by infecting the brain, for instance). Much of the early effort in chemotherapy centered on fighting these secondary infections, especially Kaposi's sarcoma (a tumor on the skin); fungal infections of the mouth; and cytomegalovirus infections of the eye, gastrointestinal tract and other organs.

Untreated, HIV infection will lead to such AIDS-related symptoms in 10 to 12 years, on average, although some patients progress much more quickly or slowly. A small minority, of great interest to researchers, appears to be able to live with the infection for an indefinite period with no ill effects at all. But untreated patients who develop full-blown AIDS generally survive only one or two more years.

The critical scientific problem is the enormous mutability of the virus. HIV is a retrovirus, which means it codes its genetic information in RNA, a nucleic acid that transfers genetic codes from DNA to proteins. The way HIV copies that information to form new viruses in the host is highly prone to error, which makes it liable to natural mutations. A patient infected with a single line of the virus may quickly develop a swarm of loosely related variants that respond differently to any treatment the patient undergoes. This variability is why HIV so readily develops resistance to drugs.

To fight this mutability, researchers have won a measure of success with an idea called combination therapy, pioneered in the 1940s for the treatment of tuberculosis and then used in the

OLIVER MECKES, MAX PLANCK INSTITUTE FOR BIOLOGY, TÜBINGEN/PHOTO RESEARCHERS, INC. (top); OLIVER MECKES, E.O.S./GELDERBLOM/PHOTO RESEARCHERS, INC. (left)

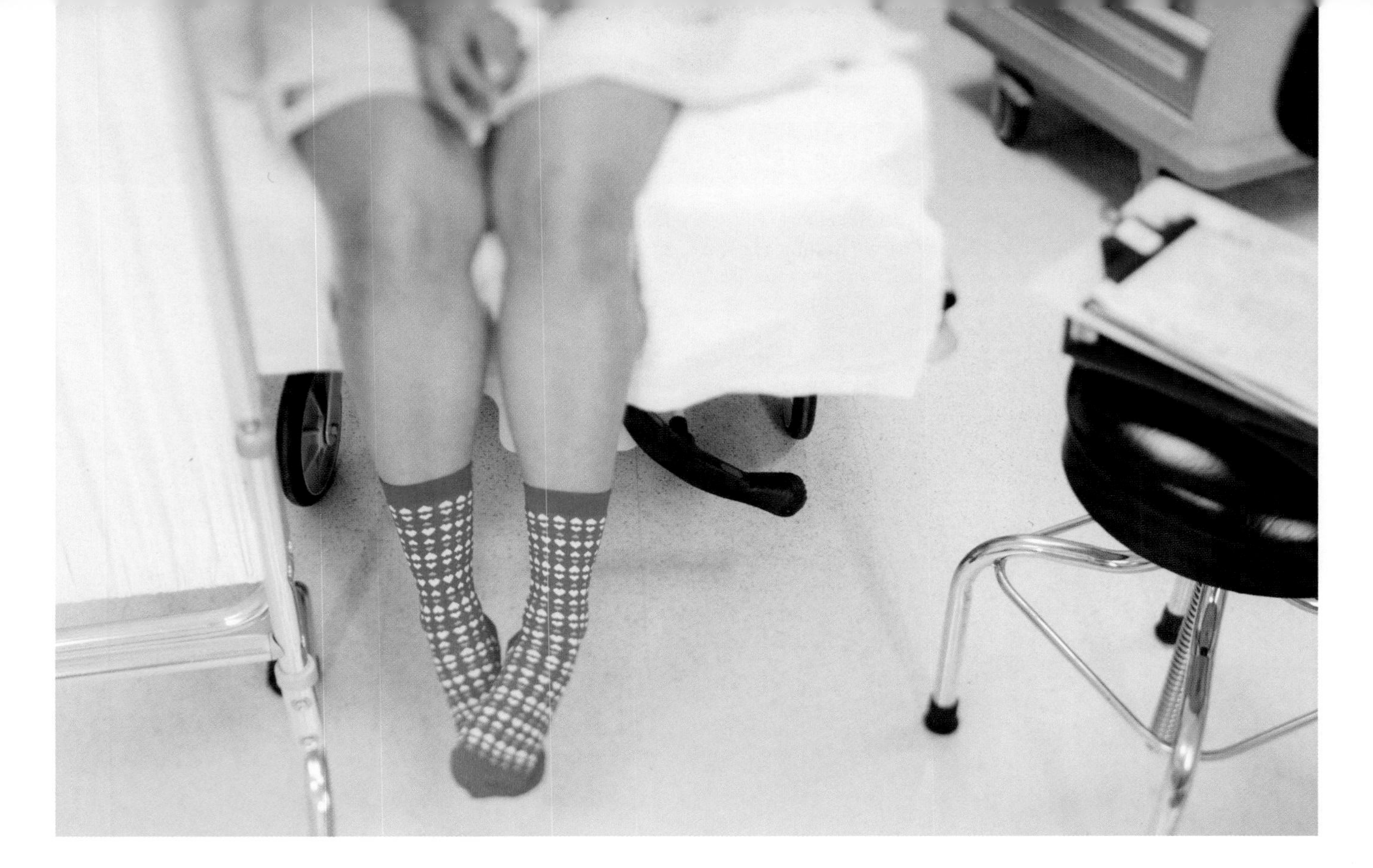

TIM PANNELL *CORBIS*

## In the case of STDs, doctor-patient confidentiality must sometimes yield to wider public-health concerns.

1970s against cancer. The idea is that a chemical "cocktail" will have an effect greater than the sum of its parts, because a mutation that saves the virus from one chemical will tend to make it vulnerable to one of the others. The cocktail's elements—called antiretrovirals—interfere with the virus's ability to reproduce inside cells. HIV patients undergoing combination therapy must take many pills a day, at set times and either before or after meals, though this regimen is being eased by the introduction of more flexible drug formulations.

Combination therapy has greatly extended the life expectancy of people with HIV, particularly (but not exclusively) when begun early in the infection. Further improvements in the regimen are in the works, most of them involving the development of drugs to attack new molecular targets in HIV. But scientists must always be on the lookout for still newer targets, lest the virus keep evolving its way to drug resistance.

There is one drawback to these improved therapies, however: because HIV is now a disease that some people can live with, many of those at risk feel that they need not change their lives greatly to avoid it. Safe-sex practices, such as the regular use of condoms, have declined, particularly in the gay community. Meanwhile, public-health authorities worry that the epidemic will continue to edge into new populations.

The CDC estimates that 40,000 new HIV infections occur in the U.S. every year, 70 percent among men and 30 percent among women. It estimates that 40 percent of these infections stem from male-to-male sex, 30 percent from male-to-female or female-to-male sex and the rest from the sharing of hypodermic needles. In the world as a whole, there are some 40 million people with HIV, a number that grows by about five million a year. Three million die of AIDS every year.

These are truly appalling numbers, both in their size and their rate of growth. Most horrifying are the statistics from Africa, which accounts for three million of the five million new cases every year, and for 560,000 of the 640,000 annual new infections of children (through congenital transmission). What is more, HIV, by weakening the immune system, is also causing flare-ups of dormant cases of tuberculosis, which then is able to spread even to people who do not have HIV. This process appears to have increased the death rate from TB in Africa by as much as 20 percent.

The economic burden is even higher than these bare numbers suggest, because HIV, like other STDs, strikes people in the prime of life. Most of the investment in their rearing, education and job training has already been made; most of the expected return of that investment has not been realized. When they die, they hollow out the demographic structure of their countries, leaving them with a higher-than-usual proportion of dependents, both young and old. In Botswana, one

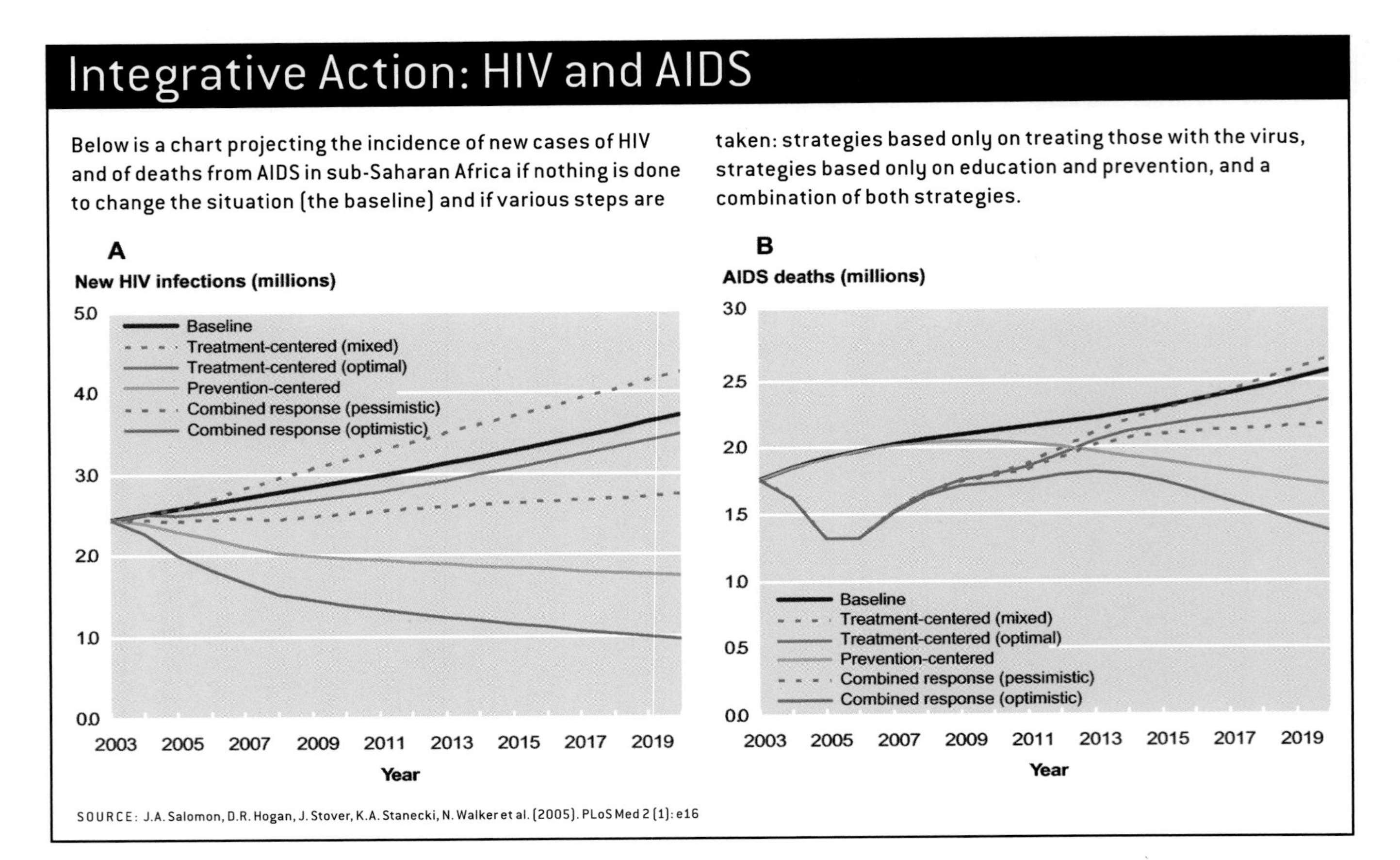


## Integrative Action: HIV and AIDS

Below is a chart projecting the incidence of new cases of HIV and of deaths from AIDS in sub-Saharan Africa if nothing is done to change the situation [the baseline] and if various steps are taken: strategies based only on treating those with the virus, strategies based only on education and prevention, and a combination of both strategies.

SOURCE: J.A. Salomon, D.R. Hogan, J. Stover, K.A. Stanecki, N. Walker et al. (2005). PLoS Med 2 (1): e16

**In Botswana, one of the hardest-hit countries of all, about 40 percent of all adults carry the HIV virus.**

of the hardest-hit countries of all, about 40 percent of all adults carry the HIV virus; the percentage of working-age adults who carry it is surely higher still.

Most people in the poor nations of the world have little access to antiretroviral therapy. What is more, most have had little education in the ways of safe sex, and those who have are often in no position to apply the knowledge. (Women in Africa, for instance, are often forced to endure unprotected sex with husbands who carry the virus.)

Discouraging as these data are, there is much that can be done. Antiretroviral therapy is increasingly being made available in poorer countries, especially for mothers giving birth, to prevent congenital transmission of HIV. Far greater gains can be expected from policies designed to spread awareness of the disease and techniques, such as condom use, that prevent it. A combination of treatment-based and prevention-based strategies appears likely to slow the infection and mortality rates most *[see chart above]*.

A complete solution to the problem of HIV, however, requires a vaccine. Even an imperfect immunization, one that lowered the chance of contracting the illness by, say, 50 percent, would be of immense benefit in slowing transmission. Unfortunately, no such vaccine is yet within reach.

### What Lies Ahead

AIDS represents a daunting challenge to global public health, and a great deal of research, energy and resources are rightly being spent to combat and eventually eradicate the disease. But there is a larger issue to consider. It would be shortsighted to regard the fight against HIV and other STDs as a mere mop-up operation, a purely political question of mustering the will to use the tools we have to eliminate these diseases in all parts of the world.

The fact is that STDs will never disappear entirely, because new ones will always be popping up. Their mode of transmission offers an ecological niche that is ideal for microbes, and the increasing interconnection of the world's populations means that our entire planet will become more and more of an incubation chamber. We must regard ourselves as surrounded by infectious agents, any one of which could emerge as the next great STD scourge. Understanding how these diseases propagate, and heeding the lessons learned from the past battles against diseases like syphilis, as well as the present fight against AIDS, will be vital in future fights against as-yet-unknown STDs.

**PHILIP E. ROSS** is a contributing editor at *Scientific American*. His work has also appeared in *IEEE Spectrum*, *Forbes*, and *Red Herring*.

# A tutor that goes where you go.

anytime.
anywhere.
anyplace.

## NEW! MP3 Tutor Sessions

Narrated by co-author Eric Simon!

Biology can be a challenging subject, but now—with new MP3 Tutor Sessions narrated by co-author Eric Simon—you've never had so much help making sense of it all! MP3 Tutor Sessions carefully walk you through the most difficult and challenging biology topics. Downloadable, portable MP3 Tutor Sessions are great for active lifestyles, a must if you are a non-native speaker or an auditory learner, and available complimentary with new text purchase at www.campbellbiology.com!
To hear a demo, just go to www.aw-bc.com/info/MP3TutorSessions.

## NEW! E-books

Now you can study wherever you have access to a computer, whether or not you have your textbook on hand! Included with new text purchase at www.campbellbiology.com!

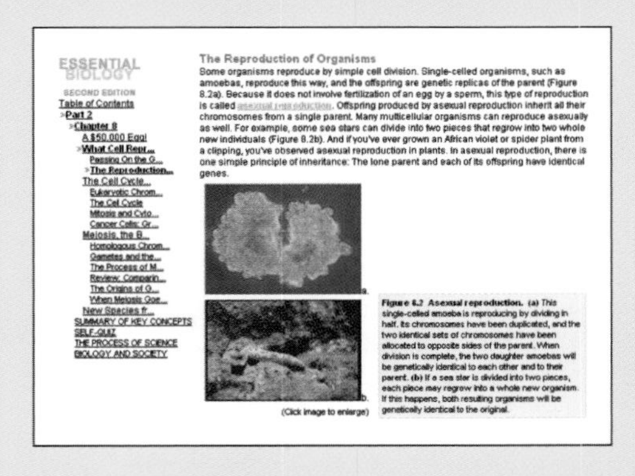

ESSENTIAL BIOLOGY
SECOND EDITION
Table of Contents
Part 2
Chapter 8
A $50,000 Egg!
What Cell Repr...
Passing On the G...
The Reproduction...
The Cell Cycle...
Eukaryotic Chrom...
The Cell Cycle
Mitosis and Cyto...
Cancer Cells: Gr...
Meiosis, the B...
Homologous Chrom...
Gametes and the...
The Process of M...
Review: Comparin...
The Origins of G...
When Meiosis Goe...
New Species fr...
SUMMARY OF KEY CONCEPTS
SELF-QUIZ
THE PROCESS OF SCIENCE
BIOLOGY AND SOCIETY

**The Reproduction of Organisms**
Some organisms reproduce by simple cell division. Single-celled organisms, such as amoebas, reproduce this way, and the offspring are genetic replicas of the parent (Figure 8.2a). Because it does not involve fertilization of an egg by a sperm, this type of reproduction is called asexual reproduction. Offspring produced by asexual reproduction inherit all their chromosomes from a single parent. Many multicellular organisms can reproduce asexually as well. For example, some sea stars can divide into two pieces that regrow into two whole new individuals (Figure 8.2b). And if you've ever grown an African violet or spider plant from a clipping, you've observed asexual reproduction in plants. In asexual reproduction, there is one simple principle of inheritance: The lone parent and each of its offspring have identical genes.

(Click image to enlarge)

**Figure 8.2 Asexual reproduction.** **(a)** This single-celled amoeba is reproducing by dividing in half. Its chromosomes have been duplicated, and the two identical sets of chromosomes have been allocated to opposite sides of the parent. When division is complete, the two daughter amoebas will be genetically identical to each other and to their parent. **(b)** If a sea star is divided into two pieces, each piece may regrow into a whole new organism. If this happens, both resulting organisms will be genetically identical to the original.

**New MP3 Tutor Sessions and E-books included with new text purchase of the following books:**

***Biology: Concepts & Connections,* Fifth Edition**
Campbell • Reece • Taylor • Simon
©2006 • 0-8053-7160-5

***Essential Biology,* Second Edition**
Campbell • Reece • Simon
©2004 • 0-8053-7473-6

***Essential Biology with Physiology***
Campbell • Reece • Simon
©2004 • 0-8053-7476-0

www.aw-bc.com

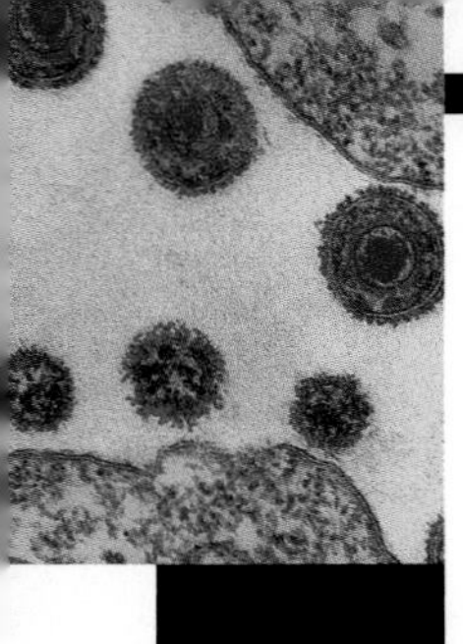

# The Challenges of STDs

*by Philip E. Ross*

# IN REVIEW

## TESTING YOUR COMPREHENSION

1) Which STD presents the greatest public health threat today?
   a) syphilis
   b) chlamydia
   c) herpes
   d) AIDS

2) If AIDS is excluded from consideration, about _____% of the burden of STD falls on women.
   a) 10
   b) 25
   c) 50
   d) 65
   e) 90

3) Vector-borne diseases tend to be _____ compared to diseases transmitted through airborne droplets.
   a) more easily cured
   b) easy to vaccinate against
   c) less severe
   d) similar in their severity
   e) more severe

4) Evolutionary medicine
   a) seeks to understand the emergence and treatment of disease based on knowledge of pathogen evolution.
   b) promotes the progressive improvement of medical practice over time.
   c) seeks to develop new approaches to disease treatment based on an evolving medical knowledge base.
   d) is devoted to the evolution of new, more natural pharmaceuticals.
   e) is the medical specialty devoted to the study and cure of STDs.

5) Educating the public to avoid more dangerous sexual practices should
   a) eliminate all STDs.
   b) result in the evolution of milder forms of STDs.
   c) reduce the number of STDs but increase their severity.
   d) reduce the number of air droplet-borne diseases.
   e) reduce the spread of vector-borne diseases.

6) Syphilis was first noted in Europe
   a) at the end of the 3rd century.
   b) in the early 9th century.
   c) at the end of the 15th century.
   d) in the early 18th century.
   e) in the late 19th century.

7) Syphilis may have evolved from bacteria that caused diseases spread by
   a) mosquitoes.
   b) the malaria parasite.
   c) sexual contact.
   d) exchange of hypodermic needles.
   e) skin-to-skin contact.

8) The most serious aspect of chlamydia infections is usually
   a) a painful, burning infection of the urethra.
   b) the high fever that often causes brain damage.
   c) the suppression of immune system function.
   d) scarring of the fallopian tubes.
   e) tumors, bone degeneration, blindness, dementia and eventually death.

9) Genital herpes is usually passed
   a) from mosquitoes to humans.
   b) from primates, especially chimpanzees, to humans.
   c) through water contaminated with fecal matter.
   d) through transfusions.
   e) from man to woman.

# ENDPOINTS

10) A major problem in combating AIDS is the
   a) high mutation rate of the HIV virus that causes the disease.
   b) extremely rapid course of the disease from infection to death.
   c) fact that infected individuals often spread the disease though casual contact.
   d) refusal of developed nations to allow the international sale of effective vaccines.
   e) fact that so little is know about how the disease spreads.

## BIOLOGY IN SOCIETY

1) When AIDS is excluded from consideration, women bear most of the burden of STDs. What aspects of STDs discussed in this article account for this uneven sharing of misery?

2) What approaches in medicine and public health education are now available to combat the spread of AIDS? Which of these approaches do you think are realistically available to a developing nation such as Rwanda with a per capita income of $237? Would the most effective practical ways to combat the AIDS epidemic be the same in a developed and a developing nation? If there is a difference, what would an effective public health policy for AIDS be in nations near the bottom and near the top of the world's economies?

3) One of the many tragedies of AIDS is that it disproportionately strikes people in the early and middle years of adulthood. In a country with a high rate of HIV infection, how does this affect the economic and social structure of the nation? If AIDS were primarily a disease of the elderly or the young, how would the economic and social impact of the disease differ?

## THINKING ABOUT SCIENCE

1) Imagine that you're a member of a medical review board considering three proposals for chemotherapeutic regimens to slow the progress of AIDS. One proposal is for treatment with a single promising new drug. The second proposal is for a new combination of three current drugs. Each of the three current drugs is known to be effective for short-term HIV chemotherapy until resistance appears and each targets a different molecule involved in HIV replication. The final proposal is also for a combination of three existing drugs, each effective for short-term treatment. The drugs in the final proposal all target the same molecule necessary for HIV replication. Which of the three proposed treatment regimens do you predict is most likely to be effective for long-term inhibition of HIV replication? Why?

2) Imagine a situation in which a bacterial pathogen that has historically been transferred between human hosts in airborne droplets acquires a mutation that allows it to replicate in mosquitoes as well as humans. Now this mutant strain can be transferred between people by a mosquito. Do you think that over a number of years the difference in the way the pathogen is spread will have any effect on the severity of the disease it causes? If there is a change, will this pathogen evolve to cause symptoms that are more or less severe? Why?

3) In the years before AIDS was understood, physicians were mystified when young patients, primarily homosexual males, were afflicted with rare bacterial and fungal infections. How does HIV infection lead to these infections? As a physician, if you diagnosed an unusual fungal infection in a young, previously healthy male, you would immediately suspect AIDS. Without running a direct test for HIV, what could you examine in this man that would strongly suggest he has AIDS?

## WRITING ABOUT SCIENCE

Since 1996, $1 billion in federal state matching funds has been committed to abstinence-only sex education programs. These programs stress abstinence from pre-marital sex and do not provide information about the relative risk of different sexual practices. Your local school board is considering instituting an abstinence-only program. Write a persuasive science-based letter to the chair of the school board arguing your position for or against abstinence-only sex education in place of a program that includes discussion of contraceptives, including condoms, and safer sex practices. Your letter should include arguments based on your knowledge of how STDs spread and your views on the likely effectiveness of an educational program to eliminate pre-marital sex in a large and diverse population.

**Testing Your Comprehension Answers:**
**1d, 2e, 3e, 4a, 5b, 6c, 7e, 8d, 9e, 10a.**

ALAN WITSCHONKE

# Taming Lupus

**Teasing out the causes of this autoimmune disorder is a daunting challenge. But the payoff should be better, more specific treatments**

**By Moncef Zouali**

A 24-year-old woman undergoes medical evaluation for kidney failure and epilepsy-like convulsions that fail to respond to antiepileptic drugs. Her most visible sign of illness, though, is a red rash extending over the bridge of her nose and onto her cheeks, in a shape resembling a butterfly.

A 63-year-old woman insists on hospitalization to determine why she is fatigued, her joints hurt, and breathing sometimes causes sharp pain. Ever since her teen years she has avoided the sun, which raises painful blistering rashes wherever her skin is unprotected.

A 20-year-old woman is surprised to learn from a routine health exam that her urine has an abnormally high protein level—a sign of disturbed kidney function. A renal biopsy reveals inflammation.

Although the symptoms vary, the underlying disease in all three patients is the same—systemic lupus erythematosus, which afflicts an estimated 1.4 million Americans, including one out of every 250 African-American women aged 18 to 65. It may disrupt almost any part of the body: skin, joints, kidneys, heart, lungs, blood vessels or brain. At times, it becomes life-threatening.

Scientists have long known that, fundamentally, lupus arises from an immunological malfunction involving antibody molecules. The healthy body produces antibodies in response to invaders, such as bacteria. These antibodies latch onto specific molecules that are sensed as foreign (antigens) on an invader and then damage the interloper directly or mark it for destruction by other parts of the immune system. In patients with lupus, however, the body produces antibodies that perceive its own molecules as foreign and then launch an attack targeted to those "self-antigens" on the body's own tissues.

LUPUS, technically lupus erythematosus, means "the red wolf." It was so named because a face rash particular to the disorder often makes people look wolflike.

Self-attack—otherwise known as autoimmunity—is thought to underpin many diseases, including type 1 diabetes, rheumatoid arthritis, multiple sclerosis and, possibly, psoriasis. Lupus, however, is at an extreme. The immune system reacts powerfully to a surprising variety of the patient's molecules, ranging from targets exposed at the surface of cells to some inside of cells to even some within a further sequestering chamber, the cell nucleus. In fact, lupus is notorious for the presence of antibodies that take aim at the patient's DNA. In the test tube, these anti-DNA "autoantibodies" can directly digest genetic material.

Until recently, researchers had little understanding of the causes of this multipronged assault. But clues from varied lines of research are beginning to clarify the underlying molecular events. The work is also probing the most basic, yet still enigmatic, facets of immune system function: the distinction of self from nonself; the maintenance of self-tolerance (nonaggression against native tissues); and the control over the intensity of every immune response. The discoveries suggest tantalizing new means of treating or even preventing not only lupus but also other autoimmune illnesses.

## Some Givens

ONE THING ABOUT LUPUS has long been clear: the autoantibodies that are its hallmark contribute to tissue damage in more than one way. In the blood, an autoantibody that recognizes a particular self-antigen can bind to that antigen, forming a so-called immune complex, which can then deposit itself in any of various tissues. Autoantibodies can also recognize self-antigens already in tissues and generate immune complexes on-site. Regardless of how the complexes accumulate, they spell trouble.

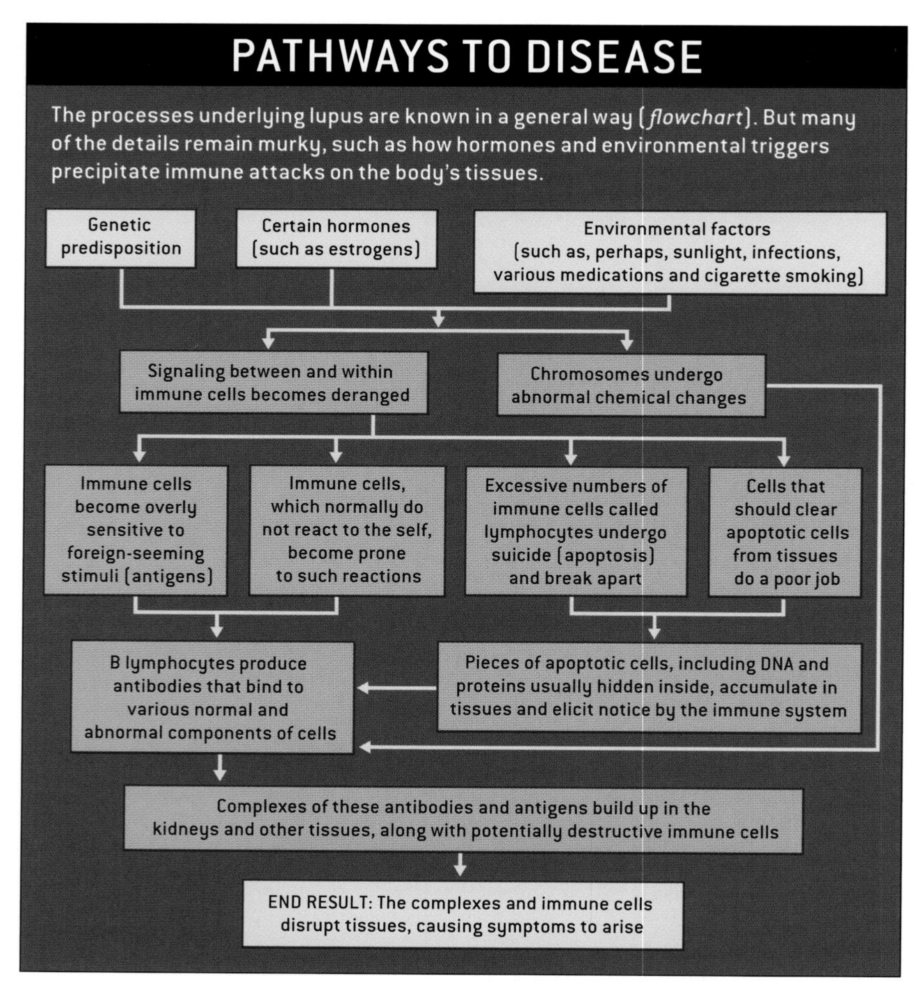


JEN CHRISTIANSEN

For one, they tend to recruit immune system entities known as complement molecules, which can directly harm tissue. The complexes, either by themselves or with the help of the complement molecules, also elicit an inflammatory response. This response involves an invasion by white blood cells that attempt to wall off and destroy any disease-causing agents. Inflammation is a protective mechanism, but if it arises in the absence of a true danger or goes on for too long, the inflammatory cells and their secretions can harm the tissues they are meant to protect. Inflammation can additionally involve the abnormal proliferation of cells native to an affected tissue, and this cellular excess can disrupt the normal functioning of the tissue. In the kidney, for instance, immune complexes can accumulate in glomeruli, the organ's blood-filtering knots of capillary loops. Excessive deposition then initiates glomerulonephritis, a local inflammatory reaction that can lead to kidney damage.

Beyond inciting inflammation, certain lupus autoantibodies do harm directly. In laboratory experiments, they have been shown to bind to and then penetrate cells. There they become potent inhibitors of cellular function.

The real mystery about lupus is what precedes such events. Genetic predisposition seems to be part of the answer, at least in some people. About 10 percent of patients have close blood relatives with the disease, a pattern that usually implies a genetic contribution. Moreover, investigators have found greater lupus concordance—either shared lupus or a shared absence of it—in sets of identical twins (who are genetically indistinguishable) than in sets of fraternal twins (whose genes generally are no more alike than those of other pairs of siblings).

## Genetic Hints

SPURRED BY such findings, geneticists are hunting for the genes at fault, including those that confer enhanced susceptibility to the vast majority of patients who have no obvious family history of the disease. Knowledge of the genes, the proteins they encode and the normal roles of these proteins should one day help clarify how lupus develops and could point to ways to better control it.

In mice prone to lupus, the work has identified more than 30 fairly broad chromosomal regions associated to some extent either with lupus or with resistance to it. Some are tied to specific elements of the disease. One region, for example, apparently harbors genes that participate in producing autoantibodies that recognize components of the cell nucleus (although the region itself does not encode antibodies); another influences the severity of the kidney inflammation triggered by lupus-related immune complexes.

In human lupus, the genetic story may be even more mind-boggling. An informative approach scans DNA from families with multiple lupus patients to identify genetic features shared by the patients but not by the other family

## Overview/*Lupus*

- Lupus arises when the immune system mistakenly produces antibodies that attack the body's own tissues, including the kidneys, skin and brain.
- The causes of this attack are complex, but a central component seems to be aberrant signaling within and between at least two types of immune cells: B lymphocytes (the antibody producers) and the T lymphocytes that help to activate the B cells.
- Several drugs under study aim to protect tissues by normalizing such signaling and quelling abnormal antibody production.

> If genes alone rarely account for the disease, **environmental contributors must play a role.**

members. Such work has revealed a connection between lupus and 48 chromosomal regions. Six of those regions (on five different chromosomes) appear to influence susceptibility most. Now investigators have to identify the lupus-related genes within those locales.

Already it seems fair to conclude that multiple human genes can confer lupus susceptibility, although each gene makes only a hard-to-detect contribution on its own. And different combinations of genes might lay the groundwork for lupus in different people. But clearly, single genes are rarely, if ever, the primary driver; if they were, many more children born to a parent with lupus would be stricken. Lupus arises in just about 5 percent of such children, and it seldom strikes in multiple generations of a family.

## Many Triggers

IF GENES ALONE rarely account for the disease, environmental contributors must play a role. Notorious among these is ultraviolet light. Some 40 to 60 percent of patients are photosensitive: exposure to sunlight, say for 10 minutes at midday in the summer, may suddenly cause a rash. Prolonged exposure may also cause flares, or increased symptoms. Precisely how it does so is still unclear. In one scenario, ultraviolet irradiation induces changes in the DNA of skin cells, rendering the DNA molecules alien (from the viewpoint of the body's immune defenses) and thus potentially antigenic. At the same time, the irradiation makes the cells prone to breakage, at which point they will release the antigens, which can then provoke an autoimmune response.

Environmental triggers of lupus also include certain medications, among them hydralazine (for controlling blood pressure) and procainamide (for irregular heartbeat). But symptoms usually fade when the drugs are discontinued. In other cases, an infection, mild or serious, may act as a lupus trigger or aggravator. One suspect is Epstein-Barr virus, perhaps best known for causing infectious mononucleosis, or "kissing disease." Even certain vaccines may provoke a lupus flare. Yet despite decades of research, no firm proof of a bacterium, virus or parasite that transmits lupus has been put forth. Other possible factors include diets high in saturated fat, pollutants, cigarette smoking, and perhaps extreme physical or psychological stress.

## Perils of Cell Suicide

ANOTHER LINE of research has revealed cellular and molecular abnormalities that could well elicit or sustain autoimmune activity. Whether these abnormalities are usually caused more by genetic inheritance or by environmental factors remains unknown. People may be affected by various combinations of influences.

One impressive abnormality involves a process known as apoptosis, or cell suicide. For the body to function properly, it has to continually eliminate cells that have reached the end of their useful life or turned dangerous. It achieves this pruning by inducing the cells to make proteins that essentially destroy the cell from within—such as by hacking to pieces cellular proteins and the chromosomes in the nucleus. But the rate of apoptosis in certain cells—notably, the B and T lymphocytes of the immune system—is excessive in those who have lupus.

When cells die by apoptosis, the body usually disposes of the remains efficiently. But in those with lupus, the disposal system seems to be defective. This double whammy of increased apoptosis and decreased clearance can promote autoimmunity in a fairly straightforward way: if the material inside the apoptotic cells is abnormal, its ejection from the cells in quantity could well elicit the production of antibodies that mistakenly perceive the aberrant material as a sign of invasion by a disease-causing agent. And such antibody production is especially likely if the ejected material, rather than being removed, accumulates enough to call attention to itself.

Unfortunately, the material that spills from apoptotic cells of those with lupus, especially the chromosomal fragments, is often abnormal. In healthy cells, certain short sequences of DNA carry methyl groups that serve as tags controlling gene activity. The DNA in circulating immune complexes from lupus patients is undermethylated. Scientists have several reasons to suspect that this methylation pattern might contribute to autoimmunity. In the test tube, abnormally methylated DNA can stimulate a number of cell types involved in immunity, including B lymphocytes, which, when mature, become antibody-spewing factories. (Perhaps the body misinterprets these improperly methylated stretches as a sign that a disease-causing agent is present and must be eliminated.) Also, certain drugs known to cause lupus symptoms lead to undermethylation of DNA in T cells, which leads to T cell autoreactivity in mice.

All in all, the findings suggest that apoptotic cells are a potential reservoir of autoantigens that are quite capable of provoking an autoantibody response. In further support of this idea, intravenous

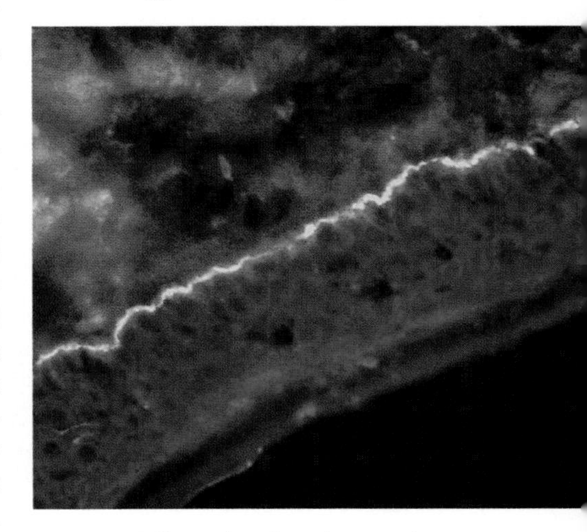

ANTIBODIES directed against tissue at the junction of the epidermis and dermis glow yellow in this micrograph of guinea pig skin exposed to blood serum from a patient with lupus. Such antibodies can trigger a damaging, inflammatory reaction.

ANDRÉ EYQUEM *Institut Pasteur, Paris*

## B CELLS GONE WRONG

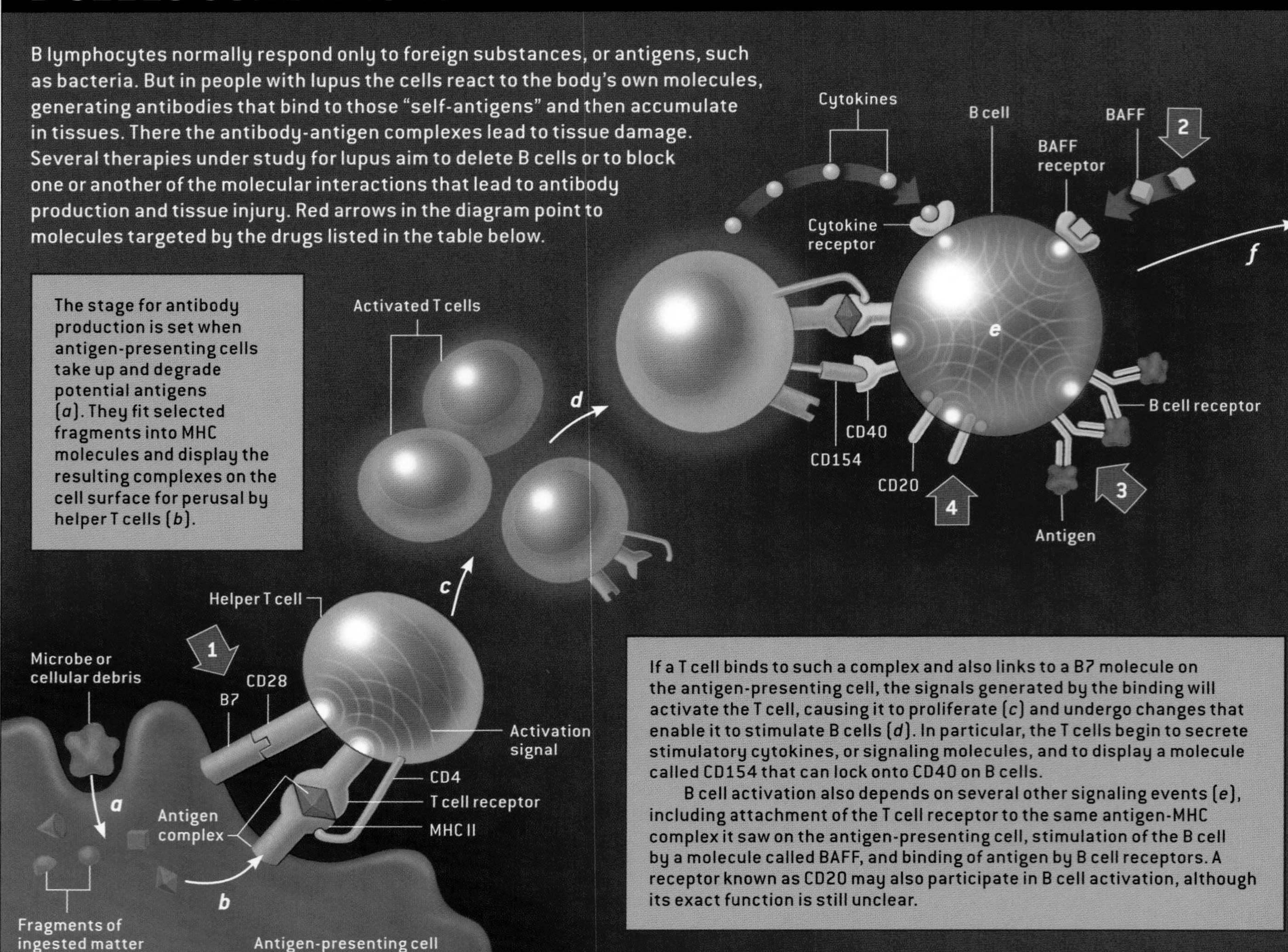


## Some Treatment Strategies under Study

| | TYPE OF AGENT | STATUS |
|---|---|---|
| 1 | **Blocker of B7's interaction with CD28, to impede activation of helper T cells** | Immune Tolerance Network, a research consortium, and the National Institutes of Health are undertaking a small human trial of a blocker called RG2077 |
| 2 | **Blocker of BAFF's interaction with its receptor, to keep BAFF (also called BLyS) from promoting B cell survival and antibody production** | Human Genome Sciences (Rockville, Md.) is evaluating one such drug, LymphoStat-B, in a multicenter trial; ZymoGenetics (Seattle) and Serono S.A. (Geneva, Switzerland) are conducting an early human trial of an agent named TACI-Ig |
| 3 | **Blocker of B cell receptors and of antibodies that recognize the body's own DNA, to inhibit the production and activity of antibodies that target such DNA** | La Jolla Pharmaceuticals (San Diego) is conducting a multicenter trial of abetimus sodium (Riquent) against lupus-related kidney disease |
| 4 | **Antibody to CD20, to deplete B cells** | Genentech (South San Francisco, Calif.) and Biogen Idec (Cambridge, Mass.) are conducting a multicenter lupus trial of rituximab (Rituxan), a drug already approved for B cell cancer |
| 5 | **Complement inhibitor, to prevent complement-mediated tissue damage** | Alexion Pharmaceuticals (Cheshire, Conn.) found evidence of disease amelioration in mice given an inhibitor of complement C5 |

JEN CHRISTIANSEN

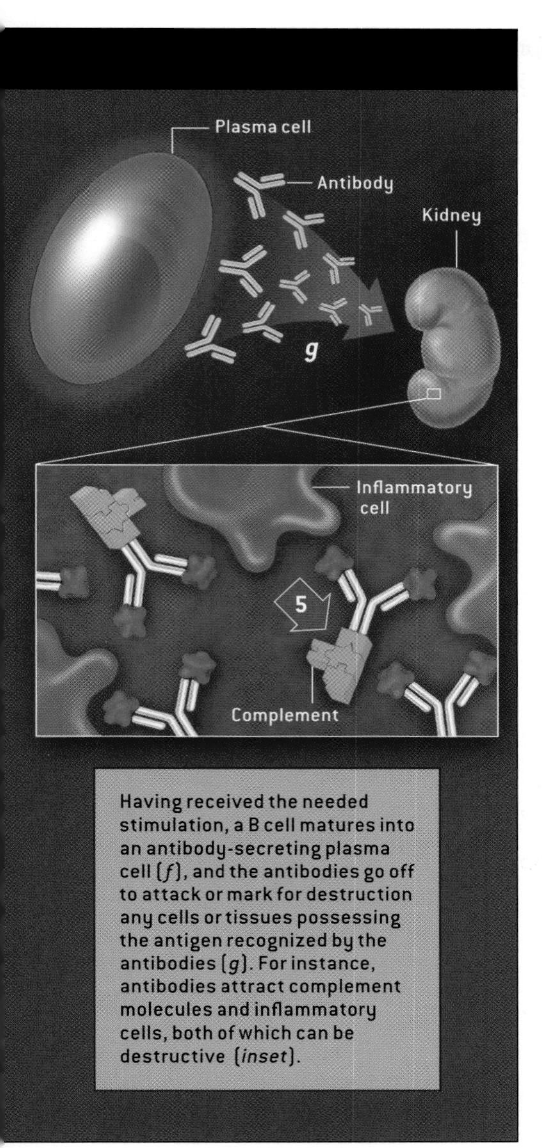


Having received the needed stimulation, a B cell matures into an antibody-secreting plasma cell (*f*), and the antibodies go off to attack or mark for destruction any cells or tissues possessing the antigen recognized by the antibodies (*g*). For instance, antibodies attract complement molecules and inflammatory cells, both of which can be destructive (*inset*).

administration of large quantities of irradiated apoptotic cells is able to induce autoantibody synthesis in normal mice.

Hence, part of the underlying process leading to the formation of destructive immune complexes may involve the body's production of foreign-seeming antigens, which cause the body to behave as if tissues bearing those antigens were alien and threatening. But other work indicates that, in addition, the B lymphocytes of lupus patients are inherently deranged; they are predisposed to generate autoantibodies even when the self-molecules they encounter are perfectly normal. In other words, the mechanisms that should ensure self-tolerance go awry.

## Deranged Cells

THE PROBLEM MOSTLY seems to stem from signaling imbalances within B cells. In the healthy body, a B cell matures into an antibody-secreting machine—known as a plasma cell—only after antibodylike projections on the B cell's surface (B cell receptors) bind to a foreign antigen. If a B cell instead attaches to a self-component, this binding normally induces the cell to kill itself, to retreat into a nonresponsive (anergic) state or to "edit" its receptors until they can no longer recognize the self-antigen.

Whether the cell responds appropriately depends in large measure on the proper activity of the internal signaling pathways that react to external inputs. Mouse studies show that even subtle signaling imbalances can predispose animals to produce antibodies against the self. And various lines of evidence indicate that certain signaling molecules (going by such names as Lyn, CD45 and SHP-1) on and in B cells of patients with lupus are present in abnormal amounts.

Other work suggests that it is not only the B cells that are deranged. For a B cell to become an antibody maker, it must do more than bind to an antigen. It must also receive certain stimulatory signals from immune system cells known as helper T lymphocytes. Helper cells of lupus patients are afflicted by signaling abnormalities reminiscent of those in the B cells. The T cell aberrations, though, may lead to autoantibody production indirectly—by causing the T cells to inappropriately stimulate self-reactive B cells.

All theorizing about the causes of lupus must account not only for the vast assortment of autoantibodies produced by patients but also for another striking aspect of the disease: the disorder is 10 times as common in women than in men. It also tends to develop earlier in women (during childbearing years). This female susceptibility—a pattern also seen in some other autoimmune diseases—may stem in part from greater immune reactivity in women. They tend to produce more antibodies and lymphocytes than males and, probably as a result, to be more resistant to infections. Among mice, moreover, females reject foreign grafts more rapidly than the males do. Perhaps not surprisingly, sex hormones seem to play a role in this increased reactivity, which could explain why, in laboratory animals, estrogens exacerbate lupus and androgens ameliorate it.

Estrogens could pump up immune reactivity in a few ways. They augment the secretion of prolactin and growth hormone, substances that contribute to the proliferation of lymphocytes, which bear receptor molecules sensitive to estrogens. Acting through such receptors, estrogens may modulate the body's immune responses and may even regulate lymphocyte development, perhaps in ways that impair tolerance of the self.

## Toward New Therapies

THOSE OF US who study the causes of lupus are still pondering how the genetic, environmental and immunological features that have been uncovered so far collaborate to cause the disease. Which events come first, which are most important, and how much do the underlying processes differ from one person to another? Nevertheless, the available clues suggest at least a partial scenario for how the disease could typically develop.

The basic idea is that genetic susceptibilities and environmental influences may share responsibility for an impairment of immune system function—more specifically an impairment of the signaling within lymphocytes and possibly within other cells of the immune system, such as those charged with removing dead cells and debris. Faulty signaling, in turn, results in impaired self-tolerance, accelerated lymphocyte death, and defective disposal of apoptotic cells and

THE AUTHOR

*MONCEF ZOUALI*, an immunologist and molecular biologist, is a director of research at INSERM, the French national institute for medical research. He focuses on conducting basic research into the molecular causes of systemic autoimmune diseases and on translating scientific insights into useful approaches to disease management. Zouali has edited several books on autoimmunity and won research awards.

## THE DIAGNOSTIC CHALLENGE

Physicians who suspect that a patient, whether female or male, has lupus continue to be hampered by the lack of a conclusive test. Because immunological self-attack may underlie many illnesses, even a classic sign of lupus—the presence of antinuclear autoantibodies—does not unmistakably diagnose this disorder.

In the absence of a sure test, doctors might gather information from a variety of sources, including not only laboratory tests but also the patient's description of current symptoms and medical history. To assist, the American College of Rheumatology has issued a list of 11 criteria that could indicate lupus. Seven concern symptoms, such as arthritis, sensitivity to sunlight or a butterfly facial rash. (The butterfly pattern is still unexplained.) The other four describe laboratory findings that include the presence of antinuclear autoantibodies or depressed concentrations of lymphocytes.

Researchers will consider a subject to have lupus if the person meets four of the criteria, but physicians might base a diagnosis on fewer cues, especially if a patient has strong indicators of the disorder, such as clinical evidence of abnormalities in several different organ systems combined with the presence of antinuclear autoantibodies. For more on common manifestations of lupus, visit the Lupus Foundation of America: www.lupus.org/ or the Lupus site: www.uklupus.co.uk/ *—M.Z.*

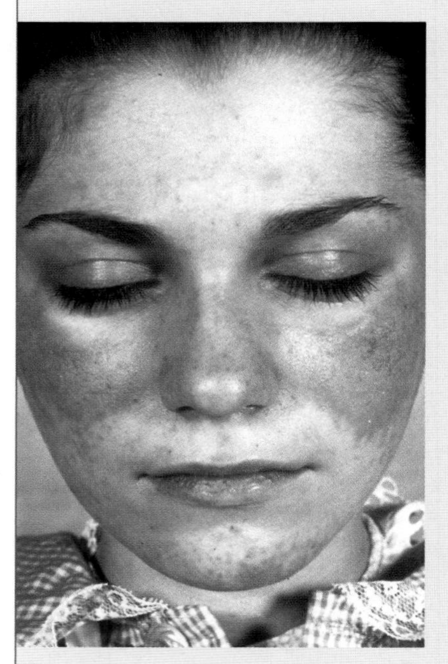

CLASSIC BUTTERFLY RASH was once thought to be the only effect of lupus.

### Current Criteria

| |
|---|
| **Malar rash** (a rash, often butterfly-shaped, over the cheeks) |
| **Discoid rash** (a type involving red raised patches) |
| **Photosensitivity** (reaction to sunlight in which a skin rash arises or worsens) |
| **Nose or mouth ulcers**, typically painless |
| **Nonerosive arthritis** (which does not involve damage to the bones around the joints) in two or more joints |
| **Inflammation of the lining in the lung or heart** (also known as pleuritis or pericarditis) |
| **Kidney disorder** marked by high levels in the urine of protein or of abnormal substances derived from red or white blood cells or kidney tubule cells |
| **Neurological disorder** marked by seizures or psychosis not explained by drugs or metabolic disturbances (such as an electrolyte imbalance) |
| **Blood disorder** characterized by abnormally low concentrations of red or white blood cells or platelets (specifically, hemolytic anemia, leukopenia, lymphopenia or thrombocytopenia) and not caused by medications |
| **Positive test for antinuclear antibodies (ANA)** not explained by drugs known to trigger their appearance |
| **Positive test for antibodies against double-stranded DNA or certain phospholipids** or a false positive result on a syphilis test |



the self-antigens they release. Abundantly available to the body's unbalanced immune surveillance, the antigens then misdirect the immune system, inducing it to attack the self.

Drugs do exist for lupus, but so far they focus on dampening the overall immune system. In other words, they are nonspecific: instead of targeting immunological events underlying lupus in particular, they dull the body's broad defenses against infectious diseases. Corticosteroids, for instance, reduce inflammation at the cost of heightening susceptibility to infections.

The challenge is to design new drugs that prevent autoimmune self-attacks without seriously hobbling the body's ability to defend itself against infection. To grasp the logic of the approaches being attempted, it helps to know a bit more about how helper T cells usually abet the transformation of B cells into vigorous antibody makers [*see box on pages 58 and 59*].

First, the helper cells themselves must be activated, which occurs through interactions with so-called professional antigen-presenting cells (such as macrophages and dendritic cells). These antigen presenters ingest bacteria, dead cells and cellular debris, chop them up, join the fragments to larger molecules (called MHC class II molecules) and display the resulting MHC-antigen complexes on the cell surface. If the receptor on a helper T cell recognizes a complex and binds to it, the binding conveys an antigen-specific signal into the T cell. If, at the same time, a certain T cell projection near the receptor links to a particular partner (known as a B7 molecule) on the antigen-presenting cell, this binding will convey an antigen-independent, or costimulatory, signal into the T cell. Having received both messages, the T cell will switch on; that is, it will produce or display molecules needed to activate B cells and will seek out those cells.

Like the professional antigen-presenting cells, B cells display fragments of ingested material—notably fragments of an antigen they have snared—on MHC class II molecules. If an activated helper T cell binds through its receptor to such a complex on a B cell, and if the T and B cells additionally signal each other through co-stimulatory surface molecules, the B cell will display receptors for small proteins called cytokines. These cytokines, which are secreted by activated helper T cells, induce the B cell to proliferate and mature into a plasma cell, which dispatches antibodies that specifically target the same antigen recognized by the coupled B and T cells.

Of course, any well-bred immune response shuts itself off when the danger has passed. Hence, after an antigen-presenting cell activates a helper T cell, the T cell also begins to display a "shutoff" switch known as CTLA-4. This molecule binds to B7 molecules on antigen-presenting cells so avidly that it links to most or even all of them, thereby putting a break on any evolving helper T and, consequently, B cell responses.

One experimental approach to treating lupus essentially mimics this shutoff step, dispatching CTLA-4 to cap over B7 molecules. In mice prone to lupus, this method prevents kidney disease from progressing and prolongs life. This substance is beginning to be tested in lupus patients; in those with psoriasis, initial clinical trials have shown that the treatment is safe.

A second approach would directly impede the signaling between helper T cells and B cells. The T cell molecule that has to "clasp hands" with a B cell molecule to send the needed co-stimulatory signal into B cells is called CD154. The helper cells of lupus patients show increased production of CD154, and in mice prone to the disease, antibodies engineered to bind to CD154 can block B cell activation, preserve kidney function and prolong life. So far early human tests of different versions of anti-CD154 antibodies have produced a mixture of good news and bad. One version significantly reduced autoantibodies in the blood, protein in the urine and certain symptoms, but it also elicited an unacceptable degree of blood-clot formation. A different version did not increase thrombosis but worked poorly. Hence, no one yet knows whether this approach to therapy will pan out.

**The challenge is to prevent immune self-attacks without hobbling the body's ability to defend itself.**

A third strategy would interfere with B cell activity in a different way. Certain factors secreted by immune system cells, such as the cytokine BAFF, promote cell survival after they bind to B cells. These molecules have been implicated in various autoimmune diseases, including lupus and its flares: mice genetically engineered to overproduce BAFF or one of its three receptors on B cells develop signs of autoimmune disease, and BAFF appears to be overabundant both in lupus-prone mice and in human patients. In theory, then, preventing BAFF from binding to its receptors should minimize antibody synthesis. Studies of animals and humans support this notion. In mice, a circulating decoy receptor, designed to mop up BAFF before it can find its true receptors, alleviates lupus and lengthens survival. Findings for a second decoy receptor are also encouraging. Human trials are in progress.

Targeting other cytokines might help as well. Elevated levels of interleukin-10 and depressed amounts of transforming growth factor beta are among the most prominent cytokine abnormalities reported in lupus, and lupus-prone mice appear to benefit from treatments that block the former or boost the latter. Taking a different tack, investigators studying various autoimmune conditions are working on therapies aimed specifically at reducing B cell numbers. An agent called rituximab, which removes B cells from circulation before they are able to secrete antibodies, has shown promise in early trials in patients with systemic lupus.

Some other therapies under investigation include molecules designed to block production of anti-DNA autoantibodies or to induce those antibodies to bind to decoy compounds that would trap them and provoke their degradation. An example of such a decoy is a complex consisting of four short DNA strands coupled to an inert backbone. Although the last idea is intriguing, I have to admit that the effects of introducing such decoys are apt to be complex.

Certain cytokines might be useful as therapies, but these and other protein drugs could be hampered by the body's readiness to degrade circulating proteins. To circumvent such problems, researchers are considering gene therapies, which would give cells the ability to produce useful proteins themselves. DNA encoding transforming growth factor beta has already been shown to treat lupus in mice, but too few tests have been done yet in humans to predict how useful the technique will be in people. Also, scientists are still struggling to perfect gene therapy techniques in general.

As treatment-oriented investigators pursue new leads for helping patients, others continue to probe the central enigmas of lupus. What causes the aberrant signaling in immune cells? And precisely how does such deranged signaling lead to autoimmunity? The answers may well be critical to finally disarming the body's mistaken attacks on itself. SA

## MORE TO EXPLORE


**Dubois' Lupus Erythematosus.** Sixth edition. Daniel J. Wallace and Bevra H. Hahn. Lippincott Williams & Wilkins, 2001.

**Immunobiology: The Immune System in Health and Disease.** Sixth edition. Charles A. Janeway, Paul Travers, Mark Walport and Mark J. Shlomchik. Garland Science, 2004.

**B Lymphocyte Signaling Pathways in Systemic Autoimmunity: Implications for Pathogenesis and Treatment.** Moncef Zouali and Gabriella Sarmay in *Arthritis & Rheumatism,* Vol. 50, No. 9, pages 2730–2741; September 2004.

**Molecular Autoimmunity.** Edited by Moncef Zouali. Springer Science and Business Media (in press).

**www.lupusresearch.org**

# Taming Lupus

***by Moncef Zouali***

# IN REVIEW

## TESTING YOUR COMPREHENSION

1) Lupus is difficult to diagnose with certainty because
   a) it is always fatal.
   b) there are many symptoms that vary between individuals.
   c) it always produces a characteristic red facial rash.
   d) its precise underlying causes are not fully known.

2) A self-antigen is
   a) a foreign substance that triggers antibody production.
   b) a foreign substance that triggers production of an antigen.
   c) a normally-occurring substance within an individual that triggers antibody production.
   d) a disease-causing antigen within an individual that triggers antibody production.

3) Genetic studies in humans have identified
   a) which gene is responsible for lupus.
   b) the region of one chromosome that contains the gene responsible for lupus.
   c) about 48 specific genes responsible for lupus.
   d) 48 broad chromosomal regions that contain a gene or genes contributing to lupus.

4) Identical twins tend to share lupus or be free of lupus more frequently than randomly selected twins. This suggests that
   a) lupus always runs in families.
   b) lupus is a purely genetic disease.
   c) lupus is not a genetic disease.
   d) lupus has a genetic component.

5) Based on current understanding, which of the following approaches would be most likely to prevent the development of lupus?
   a) Increase the occurrence of apoptosis (cellular suicide) in immune system cells and increase the disposal rate for apoptotic cells (cells undergoing suicide).
   b) Increase the occurrence of apoptosis in immune system cells and decrease the disposal rate for apoptotic cells.
   c) Decrease the occurrence of apoptosis in immune system cells and increase the disposal rate for apoptotic cells.
   d) Decrease the occurrence of apoptosis in immune system cells and decrease the disposal rate for apoptotic cells.

6) Imagine that you are a physician who is advising a patient just diagnosed with lupus. Based on current understanding of the disease, which one of the following lines of advice is justified?
   a) Think carefully about having children, as any child you have almost certainly will have lupus.
   b) Avoid aspirin or other non-steroidal anti-inflammatory medications as these aggravate lupus.
   c) Avoid sunlight because ultraviolet light often worsens lupus symptoms.
   d) Think of starting a long-term, low-dose program of antibiotics to prevent the bacterial infections known to trigger lupus symptoms.

7) A ____ cell can be converted to an antibody-producing plasma cell.
   a) B
   b) helper T lymphocyte
   c) MHC-presenting
   d) dendritic

8) The primary problem with immune system cells in lupus is
   a) a lack of B cells.
   b) an excessive numbers of T cells.
   c) deranged signaling in T and B cells.
   d) a greatly elevated mutation rate in T and B cells.

ENDPOINTS

9) Which one of the following would be an effective method for decreasing antibody production in lupus patients?
   a) Preventing B and helper T cells from binding together.
   b) Preventing B and helper T cells from coming apart.
   c) Increasing the long-term survival of B cells.
   d) Increasing the efficiency of binding between helper T cells and antigen-presenting cells.

10) Why would long-term treatment of lupus with the powerful immunosuppressant drugs available today be a problem?
   a) These drugs do not inhibit the immune system cells involved in lupus.
   b) Because a complete understanding of what goes wrong in lupus is not available, treatment with general immunosuppressant durgs would be unjustifiably dangerous.
   c) These drugs only suppress normal immunity, not autoimmunity.
   d) These drugs are blunt tools, equally suppressing beneficial and destructive immune reactions.

## BIOLOGY IN SOCIETY

1) Lupus is a relatively common disease, affecting roughly 1 in 200 Americans. The causes of lupus are complex and only partially understood. Because of this, potential treatments that are effective for some individuals may not be effective for others. Contrast this with a less common disease with a known single cause, such as cystic fibrosis. Cystic fibrosis affects about 1 in 2000 Americans. Even with the best of care, survival past 30 years is rare. Cystic fibrosis is caused by mutations in a single gene, with the majority of cases due to only a few well characterized mutations. Given the relative simplicity of cystic fibrosis, finding a cure or effective treatment for this life-threatening disease seems an easier task than for the more common but generally non-life-threatening disease of lupus. What types of questions are faced by directors of the National Institutes of Health (NIH), the primary funding agency for biomedical research in the U.S., in allocating its finite budget for research on cystic fibrosis and lupus? How would you advise the directors of NIH on the use of research dollars for these two diseases?

2) Many lupus treatments are under investigation. In those that go to clinical trial, participants will be placed into experimental and control groups. Those in control group will receive a placebo (a sham pill) instead of the experimental drug. Participants will not know which group they are in. The hope, of course, is that the experimental treatment is beneficial. Can withholding a potentially beneficial treatment from some while providing it to others within a clinical study be justified? Are there any circumstances other than potential harm by the experimental drug that might justify terminating the study earlier than planned? If so, what are these? If the clinical trial involved a life-threatening disease, such as breast cancer, would your answers change?

3) Lupus presents a significant diagnostic challenge. If you were a physician, would you prefer to err on the side of incorrectly diagnosing lupus in an individual without the disease or to miss the diagnosis of lupus in an individual who has the disorder? Which misdiagnosis would you worry about most if you thought the patient might file a medical malpractice suit?

## THINKING ABOUT SCIENCE

1) On pages 56 and 57, the author discusses the contribution of genetics to lupus. Some diseases, such as X-linked hemophilia, have a simple genetic basis: A mutation in one gene virtually guarantees the disease. How is the genetic basis of lupus different from that of X-linked hemophilia? What does it mean to say that there is a genetic component to lupus instead of saying it is a genetic disease? Propose how a genetic test could be devised once all the genes associated with lupus are known. How would the interpretation of such a genetic test for lupus differ from the interpretation of a genetic test for X-linked hemophilia?

2) Page 61 discusses some possible approaches to treating lupus. Treatment is good, but prevention is better. Thinking broadly, propose which biological process you'd have to target to prevent the development of lupus. If this could be done, are unwanted side-effects likely? What would these be and how could they be minimized? If you were thinking of applying a preventative approach to lupus, is there any way to identify susceptible individuals who would be good candidates for treatment?

## WRITING ABOUT SCIENCE

As you've discovered in this article, lupus is a complex, enigmatic disease. Draw on what you've learned about lupus research to write an essay that argues for the idea that a fundamental understanding of cellular and molecular biological processes is the foundation for rational and effective therapies. Your essay should demonstrate how advances in basic knowledge about cellular suicide and the ways B and T cells interact to control antibody production might be applied to quell the autoimmunity that underlies lupus.

**Testing Your Comprehension Answers:**
**1b, 2c, 3d, 4d, 5c, 6c, 7a, 8c, 9a, 10d.**

# Back to the Future of Cereals

Genomic studies of the world's major grain crops, together with a technology called marker-assisted breeding, could yield a new green revolution

BY STEPHEN A. GOFF AND JOHN M. SALMERON

RICE SEEDLINGS can be genetically tested for desirable traits.

For thousands of years, farmers have surveyed their fields and eyed the sky, hoping for good weather and a bumper crop. And when they found particular plants that fared well even in bad weather, were especially prolific, or resisted disease that destroyed neighboring crops, they naturally tried to capture those desirable traits by crossbreeding them into other plants. But it has always been a game of hit or miss. Unable to look inside the plants and know exactly what was producing their favorable characteristics, one could only mix and match plants and hope for the best.

Despite the method's inherent randomness, it has worked remarkably well. When our hunter-gatherer ancestors started settling down some 10,000 years ago, their development of agriculture allowed human society to undergo a population explosion. It is still expanding, demanding continual increases in agricultural productivity.

Yet 99 percent of today's agricultural production depends on only 24 different domesticated plant species. Of those, rice, wheat and corn account for most of the world's caloric intake. Each of these three extremely important cereals is already produced in amounts exceeding half a billion tons every year. To keep pace with a global population projected to reach nine billion by 2050, while maintaining our present average daily consumption of between 0.9 and 3.3 pounds of these grains per person, cereal crops will have to yield 1.5 percent more food every year and on a diminishing supply of cultivated land.

Plant scientists believe that crop yields have not yet reached their theoretical maximum, but finding ways to achieve that potential increase and to push the yield frontier still further is an ongoing international effort. Encouragingly, a new set of tools is revealing that some of the answers may be found by exploring the origins of the three major cereal crops.

MODERN CORN AND ITS ANCESTOR TEOSINTE look so dissimilar (*drawings*) that their relationship was questioned until genetic investigations confirmed it. By selectively propagating plants with desirable traits, ancient cultivators in what is now Mexico unwittingly favored certain versions of genes that control branching pattern, kernel structure and other attributes. By 4,400 years ago the teosinte cob's hard fruit case (*left photograph*) was gone and plump, modern-looking corn cobs (*right photograph*) carried the versions of the genes that control protein storage and starch quality in all domesticated corn today.

## Creating Modern Crops

MOLECULAR AND GENETIC studies are showing that wheat, rice and corn, as well as barley, millet, sorghum and other grasses, are far more interrelated than was once thought, so fresh insights into any one of these crop species can help improve the others. Further, many of these improvements may come from tapping the genetic wealth of our crops' wild ancestors by breeding useful traits back into the modern varieties.

Although the cereal crops are descendants of a common ancestral grass, they diverged from one another some 50 million to 70 million years ago, coming to inhabit geographically distinct regions of the world. Beginning around 10,000 years ago, farmers in the Mediterranean's Fertile Crescent are believed to have first domesticated wheat, and perhaps 1,000 years later, in what is now Mexico, farmers began cultivating an ancestor of modern-day corn. The ancient Chinese domesticated rice more than 8,000 years ago.

As our ancestors domesticated these plants, they were creating the crops we know now through a process very much like modern plant breeding. From the wild varieties, they selectively propagated and crossbred individual plants possessing desirable traits, such as bigger grains or larger numbers of grains. Plants that did not disperse their seeds were appealing, because harvesting their grain was easier, although this characteristic made a plant's propagation dependent on humans. Early cultivators also selected plants for their nutritional qualities, such as seeds with thin coats that could be eaten easily and maize varieties whose starch consistency best lent itself to making tortillas. In this way, crop plants became increasingly distinct from their progenitors and eventually rarely crossed with their wild versions. Corn became so dissimilar to its ancestor, teosinte, that its origin was commonly disputed until very recently [*see illustration above*].

This human modification of cereal plants through selective propagation and

## Overview/*Tapping Crops' Genetic Wealth*

- Comparing the genomes of major cereal crop species shows their close interrelationships and reveals the hand of humans in directing their evolution.
- Identifying the functions of individual plant genes allows scientists to search modern crops and their wild relatives for gene versions that confer desirable traits.
- With the desired gene as a traceable marker, traditional crossbreeding can become faster and more precise.

KAY CHERNUSH (*preceding page*); NINA FINKEL (*drawings*); JOHN DOEBLEY (*photographs*) (*this page*)

crossbreeding begun during prehistoric times has never stopped. Over the past century, crops have been selected for larger seed-bearing heads to increase their yields. These higher-yielding seed heads are heavy, so shorter plant heights were also bred into rice and wheat to prevent the plants from being bent to the ground by wind. Breeding for disease resistance, environmental stress tolerance and more efficient utilization of nitrogen fertilizers dramatically increased yields and their consistency, producing the green revolution of the 1960s. Corn's average yield per acre in the U.S., for example, has risen by nearly 400 percent since 1950.

Yet even during that boom period, plant breeders had little more to go on than the earliest crop cultivators. Most were limited to visible plant characteristics, or markers, such as seed size or plant architecture, to guide their selection of desirable lines for further propagation.

Still, studies of the genomes of cereal crops illustrate how prehistoric cultivators, by selecting for visible traits, were unwittingly selecting particular genes. For example, a group led by Svante Pääbo of the Max Planck Institute for Evolutionary Anthropology in Leipzig, Germany, analyzed the alleles, or versions, of specific genes in corn cobs recovered from sites in Mexico near the origin of corn domestication. Pääbo and his colleagues determined that by 4,400 years ago, domesticated corn already possessed genetic alleles that control the plant's branching pattern as well as aspects of protein and starch quality found in all modern corn varieties. In corn's wild relative, teosinte, these alleles occur in only 7 to 36 percent of plants, indicating that the selection pressure applied by early farmers to favor those alleles was rapid and thorough.

Indeed, working independently on different cereal crop species, breeders have been unknowingly altering them by selecting mutations in similar sets of genes. Trait mapping—narrowing the probable location of the gene underlying a trait to a particular chromosomal region, or locus—has shown that many of the changes humans have made in modern cereals map to similar loci in the genomes of related crop plants. The reason for this similarity is that the structures of these different crops' genomes are themselves so similar, despite millions of years of independent evolution separating the cereal species.

## Harvesting Genomes

A FEW THOUSAND trait-controlling loci have now been mapped in various domesticated cereals, revealing the surprising degree to which the plants' overall genetic maps have been conserved. The high degree of this correspondence, known as synteny, between genomes of all the grasses allows scientists to consider them as a single genetic system, meaning that any discoveries of genes or their function in one cereal crop could help scientists to understand and improve the others.

Rice, whose formal name is *Oryza sativa*, is likely to be the first to yield many of these new insights, because it will be the first crop plant to have its entire genome sequenced. One of us (Goff) has already published a draft sequence of the *japonica* subspecies of rice most commonly grown in Japan and the U.S., and Chinese researchers have produced a draft of the *indica* subspecies widely cultivated in Asia. The International Rice Genome Sequencing Project is expected to complete a detailed sequence of rice's 12 chromosomes by the end of this year.

The rice genome is the easiest of all the cereals' to tackle because it is much smaller than the others, with only 430 million pairs of DNA nucleotides. By comparison, the human genome has three billion of these so-called base pairs, as does corn. Barley's genome contains five billion base pairs and wheat, a whopping 16 billion. A corn genome–sequencing project is under way, and one for wheat is under consideration. And from the existing sequence information about rice, tens of thousands of genes have already been identified. Just knowing that a stretch of the genome is a gene does not tell us what it does, though.

Several strategies allow us to determine a gene's function, but the most straightforward involves searching existing databases of all known genes to look for a match. Often genes are responsible for such basic cellular activities that a nearly identical gene will be found in microbes or other organisms whose genes have already been studied. Of the 30,000 to 50,000 predicted genes in rice, approximately 20,000 have sequence similarity, or homology, to previously discovered genes whose function is known,

## DESIRABLE TRAITS

Traits that plant breeders seek to modify fall into broad categories, including growth, plant architecture, stress tolerance and nutrient content. Yield increases—the holy grail of agriculture—can be achieved by expanding the size or number of grains produced by a single plant, by enabling more plants to grow in the space usually needed for one, or by making plants tolerant of conditions where they previously could not thrive.

| Growth | Architecture | Stress tolerance | Nutrient content/quality |
|---|---|---|---|
| Grain size or number | Height | Drought | Starch |
| Seed-head size | Branching | Pests | Proteins |
| Maturation speed | Flowering | Disease | Lipids |
| | | Herbicides | Vitamins |
| | | Intensive fertilization | |

DOUG WILSON *Corbis*

## MATCHING TRAITS TO GENES

The same tools that allow scientists to trace some human diseases to individual genes make it possible to find the genes responsible for plant attributes. Mapping techniques can narrow the trait-controlling gene's probable location to one region on a chromosome; sequencing of the DNA in that region will then narrow the search to a likely gene. To find out the gene's function, investigators can apply any of the techniques below.

### DATABASE SEARCH

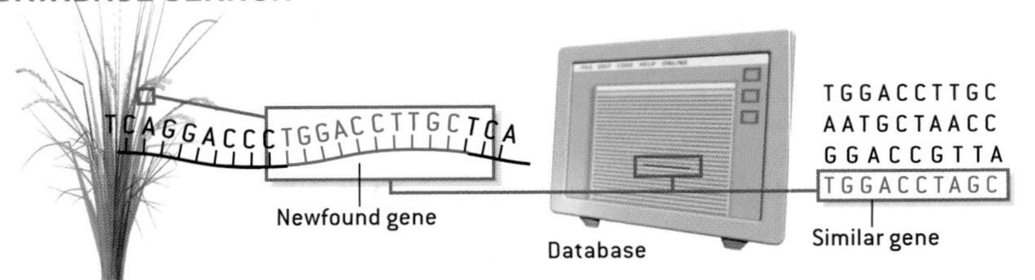


Comparing a newfound gene with known genes in a variety of databases can yield a near match. Of rice's estimated 30,000 to 50,000 genes, 20,000 are similar to genes already studied in other organisms and are assumed to have the same functions.

### EXPRESSION PROFILE

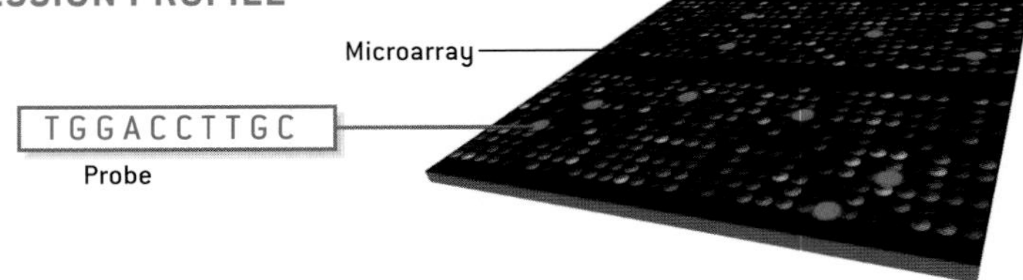


Expression profiling gives clues to a gene's function by showing when and where the gene is activated in a plant. A microarray holds thousands of snippets of DNA called probes. Each probe matches a unique signature of gene activity called a messenger RNA (mRNA). When plant cell samples are washed across the microarray, any mRNAs present will stick to their matching probes, causing the probes to emit light. If a gene is expressed, or activated, only during grain development, for example, it is assumed to play a role in that process.

### MUTANT LIBRARY

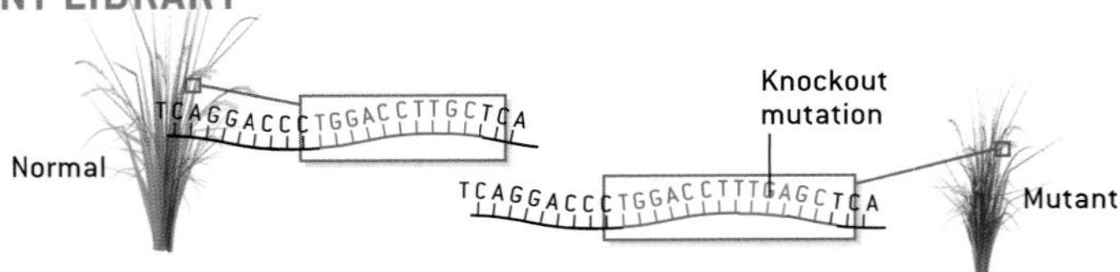


Studying mutants can reveal the function of specific genes by showing what happens when the genes are deactivated. A small piece of DNA inserted into a gene of interest can "knock out," or silence, that gene in the developing plant. Screening the mutant for physical or chemical differences from normal plants can indicate the gene's usual role.

### PREDICTED RICE GENE CLASSIFICATIONS

Using the methods described above, investigators have determined or predicted the functions of a large fraction of rice genes.

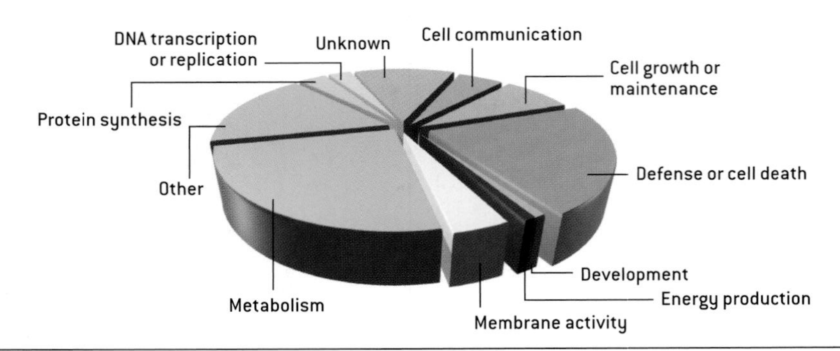


which allows researchers to predict the role of those genes in rice.

For example, more than 1,000 genes are predicted to be involved in defending rice against pathogens and pests. Likewise, hundreds of genes have been assigned to specific metabolic pathways that lead to the synthesis of vitamins, carbohydrates, lipids, proteins or other nutrients of interest. From experimental data about well-studied plants such as *Arabidopsis* (thale cress), many of the genes that regulate these biosynthetic pathways or affect important stages of crop development, such as flower and seed formation, have also been identified.

A number of research groups have gone further and begun using powerful tools called microarrays to catalogue which genes are expressed, or activated, in a variety of distinct cereal tissues. For example, scientists at our company, Syngenta, examined 21,000 rice genes and identified 269 of them that are preferentially expressed during development of the rice grain, suggesting that these genes play key roles in determining the nutrient composition of the mature grain.

A somewhat different approach to determining a gene's function is to "knock it out" by inserting a mutation into the gene that shuts off its activity and then see what happens to the plant. Sometimes the effect is visible, but the modified plant can also be tested for less obvious changes in any of its normal physiological, developmental, internal regulatory or biochemical functions. Both private and public efforts have completed collections of mutant rice and corn plants in which thousands of specific genes have been knocked out. Such functional genomic studies, combined with sequence comparisons of genes across species, allow scientists to begin developing a basic understanding of how many and which of rice's genes—and by extension those of corn, wheat, sorghum and other cereal crops—contribute to plant development, physiology, metabolism and yield.

Once the function of a specific gene is known, a remaining step in using that knowledge to improve crops is to identify specific alleles of the gene that deliver desirable traits. For example, if a gene is

SLIM FILMS

known to control an aspect of starch accumulation in corn grain, a version of the gene can be sought that functions under severe drought conditions. Such desirable alleles may be found in other modern corn varieties, but even more will probably be discovered in wild relatives of crop plants. Genetic homogeneity among modern crops is an adverse consequence of the way our ancestors initially domesticated them. According to one estimate, modern corn's founding population may have comprised as few as 20 plants. By selecting only a few individual plants with desirable traits to propagate and then inbreeding these for thousands of years, early cultivators severely limited genetic diversity in the domesticated species.

Experimenting with both tomato and rice plants, Steven Tanksley and Susan R. McCouch of Cornell University have pioneered searches for beneficial alleles in wild varieties that might improve modern crops. Their work has demonstrated the genetic diversity available in wild relatives of domesticated plants, at the same time showing that the wild varieties' most valuable resources are not always obvious. In one experiment during the mid-1990s, Tanksley crossed a tiny wild green tomato species from Peru with a somewhat pale red modern processing tomato cultivar. Surprisingly, he found that a gene from the green tomato made the red tomato redder. As it turned out, the green tomato lacked certain genes to complete synthesis of the pigment lycopene, which gives tomatoes their red hue, but it did possess a superior allele for a gene that plays a role earlier in the tomato's lycopene synthesis pathway.

The genetic variety in wild relatives of our modern crops is only beginning to be explored. In rice and tomatoes, an estimated 80 percent of each species' total allelic diversity remains untapped. Remarkable studies by Tanksley, McCouch and others have repeatedly demonstrated the ability of wild alleles to produce dramatic changes in physical aspects of domesticated plants, even though some of the changes seem counter to the wild plants' normal attributes, as in the tomato example. So without the technology to use genes or chromosomal loci as molecular markers, scientists will find identifying some of these desirable traits or moving them into modern crops nearly impossible.

AUTHORS JOHN SALMERON (*left*) AND STEPHEN GOFF with experimental corn plants. In the greenhouse, female reproductive parts of the corn plants, called silks, are covered with small white waxed-paper bags to prevent their fertilization by the male reproductive parts, or tassels, at the top of the plant.

## Marker-Assisted Breeding

ONCE SCIENTISTS HAVE identified specific sets of beneficial alleles from different wild or modern plant varieties, the goal becomes moving just those alleles into a modern crop breeding line, known as an elite cultivar. We could use bacterial DNA or some other delivery vehicle to transfer selected genes, applying the same process (called transformation) used to create so-called genetically modified crops. But scientists are also exploring an approach that avoids the long and expensive regulatory approval process for transgenic plants: breeding guided by genetic markers.

Knowing the exact alleles that confer desirable traits, or even just their chromosomal loci, a breeder could "design" a new plant that combines those traits with the best qualities of an elite cultivar, then build it through crossbreeding with the help of DNA-fingerprinting technology such as that used to determine paternity or solve forensic questions [*see illustration on next page*].

All large-scale plant breeding produces tens of thousands of seedlings. But instead of having to plant each of these progeny and wait until they mature to see if a trait has been inherited, a breeder would simply sample a bit of each seedling's DNA and scan its genes for the chosen allele, which serves as a marker for the desirable trait.

Seedlings possessing the desired allele would be grown until they were ready to crossbreed with the elite cultivar. Those progeny would then be tested for the allele, and so on, until the breeder had a population of plants that resembled the original elite cultivar but for the presence in each of a newly acquired allele. The time savings afforded by using genetic fin-

THE AUTHORS

*STEPHEN A. GOFF* and *JOHN M. SALMERON* are plant geneticists at Syngenta Biotechnology, Inc., in Research Triangle Park, N.C. Goff led a U.S. team in producing a draft sequence of the rice genome published in 2002. He is currently working on a humanitarian initiative to use genetic information about rice to help crop improvement efforts in developing countries. Salmeron, director of applied trait genetics for SBI, has been applying genetics to crop improvement since 1989, when as a postdoctoral researcher at the University of California, Berkeley, he isolated one of the first plant disease-resistance genes in the tomato.

# DESIGNING AND BUILDING NEW CROPS

When scientists know which gene controls a specific plant trait, such as seed size, they can search different varieties of the domesticated plant and its wild relatives to find a preferable version, or allele, of the gene. A breeder could then move a desirable allele from one plant into another through conventional crossbreeding, using the allele itself as a traceable marker for the trait. Instead of waiting a full growing season for plants to mature, the breeder could rapidly find out if seedlings have the desired trait by testing them for the allele in each round of breeding. Such marker-assisted breeding would dramatically shorten the time required to develop a new crop variety.

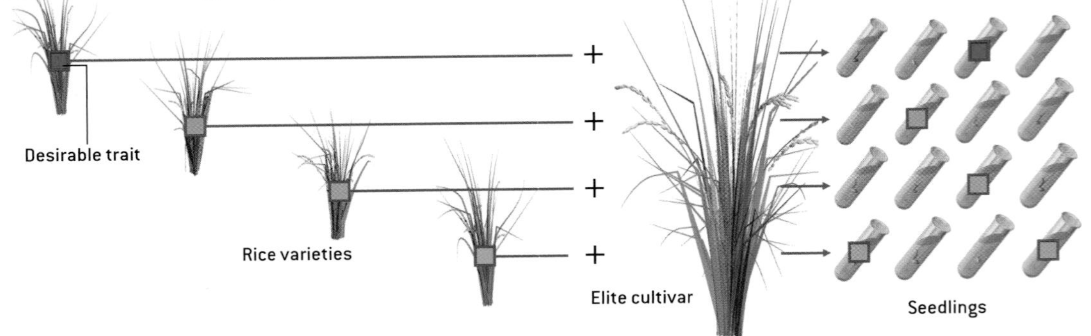


**1** Each of four different rice varieties with a desirable trait can be crossed with an elite breeding line, or cultivar, to produce tens of thousands of seedlings.

**2** Some, but not all, of the seedlings will inherit the desirable allele.

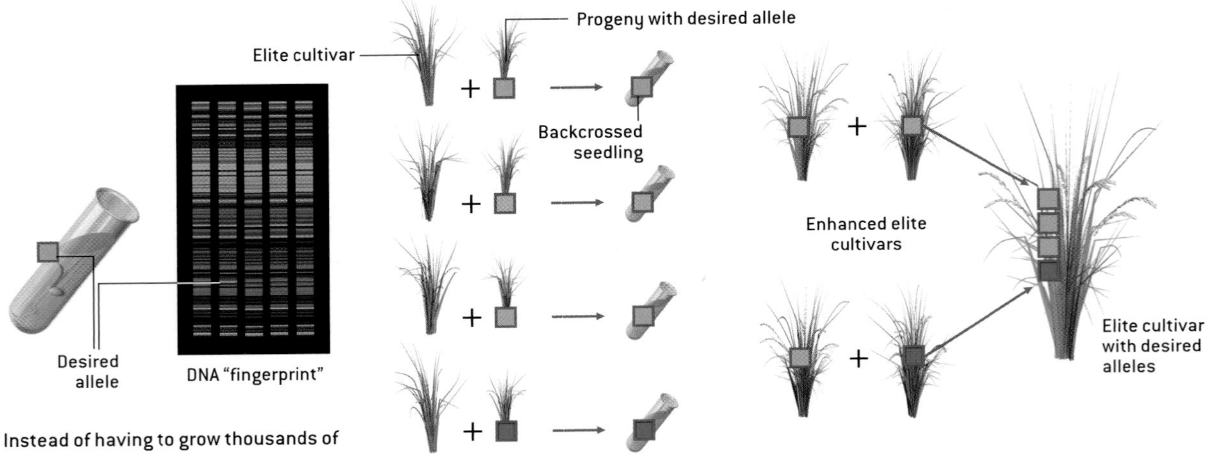


**3** Instead of having to grow thousands of plants to maturity to see which ones inherited the trait, breeders can test each seedling's DNA for the desired allele just days after germination with the technology used for so-called DNA fingerprinting.

**4** Only progeny with the desired alleles are grown until they are mature enough to breed with the elite cultivar, a step known as backcrossing.

**5** Crossing and backcrossing are repeated, with the progeny's genes tested in every round, until all the desired alleles have been moved into the elite crop plant.

## GENETIC DIVERSITY IN RICE

After thousands of years of inbreeding, modern crop varieties are far less genetically diverse than their wild relatives (*pie chart*), making the wild plants a rich reservoir for novel alleles. The untapped wealth in wild plants is not always obvious: in experiments with rice ancestor *Oryza rufipogon* (*left*), alleles from the wild plant were moved into a modern high-yield Chinese rice variety (*right*) using marker-assisted breeding. The low-yield wild plant's genes raised the modern variety's yield by 17 to 18 percent.

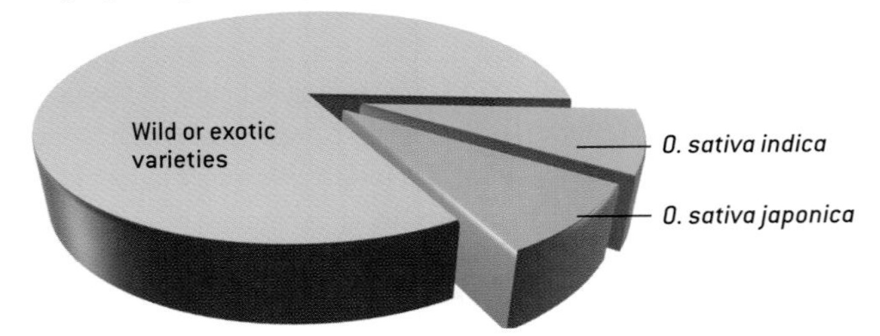

*O. rufipogon* — High-yield variety

## CEREAL CROP YIELDS

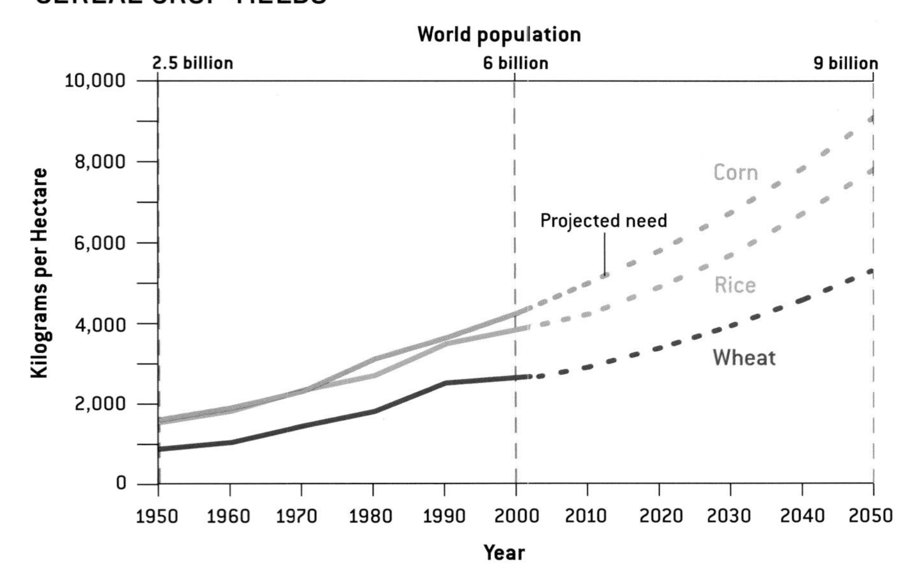


WORLDWIDE AVERAGE YIELDS for corn, rice and wheat nearly tripled between 1950 and 2000, a period that saw global population do the same. To feed a projected world population of nine billion in the year 2050, while maintaining present average consumption of 0.9 to 3.3 pounds per person a day of these cereal crops, yields must keep rising by 1.5 percent a year.

gerprinting to test for trait markers at each stage of this process would literally shave years off the time typically required to develop new crop varieties. This acceleration would allow breeders to become more responsive to changing circumstances, such as the emergence of new pests or of resistance among old pests to current countermeasures. Tailoring new crop varieties with combinations of characteristics that are optimized for different environments, farmers' needs or consumer preferences would also become easier.

But the real revolutionary potential in this method lies in its power to open up the genetic bottleneck created thousands of years ago when our major crops were first domesticated. Once scientists accumulate more information about the functions of genes in the grasses, we can more effectively search the huge reservoir of genetic diversity waiting in the wild relatives of modern crop plants. One of McCouch's experiments provides an example of the possibilities: she used molecular markers to identify gene loci in a Malaysian wild ancestor of rice known as *Oryza rufipogon*. McCouch and her colleagues then employed marker-assisted breeding to move a total of 2,000 genes—approximately 5 percent of the rice genome—into plants of a modern Chinese hybrid rice variety.

This experiment was focused on finding alleles to increase the already high-yield hybrid's output still further, and the resulting test plants were examined for several yield-improving traits, such as plant height, length of its flowering head (panicle), and grain weight. Approximately half the wild relative's loci turned out to have yield-improving alleles, although some of these also had negative effects on other aspects of the plants' growth, such as slowing maturation time. But two of the alleles from *O. rufipogon* seemed to have no negative effects and produced yield increases of 17 and 18 percent, respectively, in the modern cultivar. As in Tanksley's tomato experiment, nothing about the wild plant's appearance [*see box on opposite page*] suggested that it could teach modern rice something about yield, yet the results were impressive and encouraging.

Of course, certain beneficial genes cannot be moved into modern crop varieties by means of traditional breeding. For example, genes conferring some types of herbicide tolerance or insect resistance do not exist in plants that will crossbreed with corn.

A gene can be transferred into a recipient plant using current transformation techniques, but they do not allow scientists to specify where in the recipient organism's genome a new gene is inserted. Thus, one could add a new allele but not necessarily succeed in replacing the old, less desirable allele. Yet in the cells of mice and in some microbes, a phenomenon called homologous recombination directs an introduced gene to a chromosomal location whose DNA sequence is most similar to it—permitting a desirable allele of a gene to directly replace the original.

In the future, we may be able to achieve the same one-step substitution of gene alleles in crop plants. Homologous recombination was recently demonstrated in rice, and a related process has been used to replace alleles in corn. Once it becomes routine, the ability to swap pieces of chromosomes this way in the laboratory might allow scientists to exchange alleles between some plants that cannot be crossbred naturally.

Today marker-assisted breeding is already speeding the process in crops of the same species or close relatives. No new cereals have been domesticated in more than 3,000 years, suggesting that we will need to rely on improving our current major crops to meet ever increasing food demands. By providing the ability to peer into plants' genomes and the tools to harvest their hidden treasures, genetic science is opening the way to a new green revolution. SA

### MORE TO EXPLORE

**Seed Banks and Molecular Maps: Unlocking Genetic Potential from the Wild.** Steven D. Tanksley and Susan R. McCouch in *Science*, Vol. 277, pages 1063–1066; August 22, 1997.

**A Draft Sequence of the Rice Genome (*Oryza sativa* L. ssp. *japonica*).** Stephen A. Goff et al. in *Science*, Vol. 296, pages 92–100; April 5, 2002.

**Gramene, a Tool for Grass Genomics.** D. H. Ware et al. in *Plant Physiology*, Vol. 130, No. 4, pages 1606–1613; December 2002.

**Early Allelic Selection in Maize as Revealed by Ancient DNA.** Viviane Jaenicke-Després et al. in *Science*, Vol. 302, pages 1206–1208; November 14, 2003.

NINA FINKEL; SOURCE: FOOD AND AGRICULTURE ASSOCIATION

# Back to the Future of Cereals

***by Stephen A. Goff and John M. Salmeron***

# IN REVIEW

## TESTING YOUR COMPREHENSION

1) The first crops were cultivated about____ years ago.
   a) 3,000
   b) 10,000
   c) 50,000
   d) 100,000

2) To feed the world's growing population, crop yields must be increased____ percent each year.
   a) 0.01
   b) 0.1
   c) 1.5
   d) 10

3) The yield of corn per acre has increased____ percent in the U.S. since 1950.
   a) 5
   b) 20
   c) 100
   d) 400

4) Narrowing the probable location of a trait to a particular chromosomal region is called
   a) trait mapping.
   b) gene cloning.
   c) plant breeding.
   d) expression profile screening.

5) The____ genome was the first completely sequenced genome of a crop plant.
   a) corn
   b) wheat
   c) sorghum
   d) rice

6) Knowledge of when and where a plant gene is activated can be obtained by
   a) expression profiling.
   b) marker-assisted breeding.
   c) conventional breeding.
   d) creation of a mutant library.

7) In comparison to their wild relatives, modern crop plants have
   a) low yields.
   b) low genetic diversity.
   c) more genes.
   d) few genetic markers.

8) DNA fingerprinting is an essential tool for
   a) plant cultivation.
   b) analysis of microarrays.
   c) marker-assisted breeding.
   d) genome sequencing.

9) When a hybrid plant produced by crossing two different varieties is crossed to one of its parents, this is
   a) a backcross.
   b) marker-assisted breeding.
   c) homologous recombination.
   d) genetic engineering.

10) The method that allows researchers to tailor cultivars so they contain only desired alleles is
   a) marker-assisted breeding.
   b) trait mapping.
   c) homology searching.
   d) expression profiling.